THE LANGUAGE AND REALITY OF TIME

The Language and Reality of Time

THOMAS SATTIG

CLARENDON PRESS · OXFORD

OXFORD
UNIVERSITY PRESS

Great Clarendon Street, Oxford OX2 6DP

Oxford University Press is a department of the University of Oxford.
It furthers the University's objective of excellence in research, scholarship,
and education by publishing worldwide in

Oxford New York

Auckland Cape Town Dar es Salaam Hong Kong Karachi
Kuala Lumpur Madrid Melbourne Mexico City Nairobi
New Delhi Shanghai Taipei Toronto

With offices in

Argentina Austria Brazil Chile Czech Republic France Greece
Guatemala Hungary Italy Japan Poland Portugal Singapore
South Korea Switzerland Thailand Turkey Ukraine Vietnam

Oxford is a registered trademark of Oxford University Press
in the UK and in certain other countries

Published in the United States
by Oxford University Press Inc., New York

First published 2006

British Library Cataloguing in Publication Data
Data available

Library of Congress Cataloging in Publication Data
Sattig, Thomas.
The language and reality of time / Thomas Sattig.
p. cm.
Includes bibliographical references and index.
1. Time. I. Title.
BD638.S29 2006 115—dc22 2006000592

Typeset by Laserwords Private Limited, Chennai, India
Printed in Great Britain
on acid-free paper by
Biddles Ltd., King's Lynn, Norfolk

ISBN 0-19-927952-7 978-0-19-927952-4

1 3 5 7 9 10 8 6 4 2

For my parents,
Christel and Bernd Sattig

Acknowledgements

Most of this book was written while I was a British Academy Postdoctoral Fellow and a Junior Research Fellow of Brasenose College, Oxford. I thank the British Academy and the Principal and Fellows of Brasenose College for their support.

For valuable comments on the material presented in this book I am grateful to a number of people. I am most indebted to Tim Williamson, who supervised the doctoral thesis that forms the origin of this book, and who accompanied the book's progress with stimulating insights. I am also grateful to Jeremy Butterfield, Kit Fine, Graeme Forbes, Jonathan Lowe, Kris McDaniel, Hugh Mellor, Adrian Moore, Calvin Normore, Josh Parsons, Oliver Pooley, Gonzalo Rodriguez-Pereyra, Nick Shea, Nico Silins, Barry Smith, Stephen G. Williams, Dean Zimmerman, and participants in graduate classes I gave with Jeremy Butterfield at Oxford University. I am especially grateful to Thomas Crisp, Heather Dyke, and Ted Sider for extensive comments on the entire manuscript. Further, I thank Peter Momtchiloff for his interest and encouragement, and Ahlie Schaubel for being there.

Section 5.5 draws on material previously published in my 'Temporal Predication with Temporal Parts and Temporal Counterparts', *Australasian Journal of Philosophy*, 81 (2003), 355–68, by permission of Oxford University Press. Section 6.3 contains material from my 'Temporal Parts and Complex Predicates', *Proceedings of the Aristotelian Society*, 102 (2002), 279–86, by permission of the Editor of the Aristotelian Society Proceedings.

T.S.

Contents

Introduction

Both ordinary language and the material world have a temporal dimension. Ordinary language has a temporal dimension in that it is temporally modified; when we say that something is the case, we also indicate at what time it is the case. The material world has a temporal dimension in that it is constituted by objects that are in time. In the philosophical tradition there has been a tendency to study the temporal dimensions of language and reality separately. My project is to explore the temporal dimension of the world around us in relation to the temporal dimension of our discourse about the world.

Chapter 1 serves as an introduction to the themes of the book. I state the problem of *temporal supervenience*, which links the project of exploring the temporal dimension of language with the project of exploring the temporal dimension of reality. Time can be viewed from different angles. On one conception, time is what I call *ordinary time*. Ordinary time is an entity that has one dimension, is distinct from three-dimensional space, and consists of past, present, and future. This conception of time is 'ordinary' in virtue of being the conception that we are committed to by our ordinary temporal discourse. According to another conception, there is no one-dimensional time distinct from a three-dimensional space, but rather only a four-dimensional *spacetime* of which time is merely an aspect. Spacetime consists of a manifold of spacetime points that stand in certain temporal and spatial relations to each other. These two conceptions of time are not rivals. They are compatible conceptions serving different purposes. How is what goes on in ordinary time related to what goes on in spacetime? I find it overwhelmingly plausible that all facts about ordinary time logically supervene on facts about spacetime; what goes on in spacetime fully determines what goes on in ordinary time. This is the general thesis of temporal supervenience. The problem of temporal supervenience is to specify the facts about spacetime on which facts about ordinary time supervene, and to explain how they supervene.

Temporal supervenience has many aspects, corresponding to various kinds of supervenient temporal phenomena. Among the most basic phenomena are the following: ordinary objects, such as persons and tables, persist through ordinary time—they exist at various times; and ordinary properties, such as shapes, are instantiated at various times—if incompatible properties are instantiated by the same object at different times, then the object changes through ordinary time. The problem of temporal supervenience with respect to these phenomena has two components. The first component is to specify the spatiotemporal supervenience base of temporal existence and persistence, and of temporal instantiation and change. How do objects occupy spacetime? And how are properties instantiated across occupied spacetime? The second component is to build an explanatory bridge from the supervenience base to the supervenient phenomena. Such a bridge requires an 'analysis' of temporal existence and temporal instantiation—that is, a semantic account of ordinary temporal predications such as '*a* was F'. The problem of temporal supervenience thus connects the metaphysics of time with the semantics of temporal discourse.

Before the supervenience of ordinary temporal facts on spacetime facts can be explained, the shape of ordinary time needs to be clarified. This is a further task of Chapter 1. Since ordinary time is the conception of time to which we are committed in virtue of the way we speak, the metaphysical question of the shape of ordinary time is closely linked with the semantic question of the status of grammatical tense. The construal of ordinary time as *A-time* corresponds to the *tenserist* account of tense, whereas the construal of ordinary time as *B-time* corresponds to the *detenserist* account of tense. Tensers hold that grammatical tense is semantically irreducible, while detensers hold that tense is semantically reducible. I criticize tenserism and A-time in the context of temporal supervenience with the aim of promoting detenserism as the correct account of tense and B-time as the true shape of ordinary time. With detenserism in the background the problem of temporal supervenience becomes the task of explaining how facts about B-time supervene on facts about spacetime.

Part of the problem of temporal supervenience is the problem of spatiotemporal location: how are objects located in spacetime? In Chapter 2, I provide a detailed statement of various answers to this problem. Knowing the possible forms of spatiotemporal location is crucial for structuring the discussion of the problem of temporal supervenience. The main answers to the problem of spatiotemporal location are *three-dimensionalism* and *four-dimensionalism*. The three-dimensionalist holds that an object occupies many temporally unextended regions of spacetime, whereas the four-dimensionalist holds that an object occupies only a single temporally extended region of spacetime. Subsequently to stating these accounts of spatiotemporal location, I discuss the relationship of three-dimensionalism and four-dimensionalism to other theses and theories, including the theory of temporal parts, endurantism, perdurantism, eternalism, and presentism.

An account of temporal supervenience requires an account of temporal predication—a semantic account of the language in which facts about ordinary time are stated. For the detenser, the problem of temporal predication is essentially the task of giving an account of the semantic function of the modifier 'at *t*' in '*a* is F at *t*'. In the project of explaining temporal supervenience, an account of temporal predication functions as an analysis of ordinary temporal facts, which is required to build an explanatory bridge from these temporal facts to their spatiotemporal supervenience base. In Chapter 3, I discuss various accounts of temporal predication that share the common feature that temporal supervenience cannot be explained on the basis of them, because these accounts allow no plausible explanatory link between the facts of persistence and change and any facts about spacetime.

In Chapter 4, a new account of temporal predication—the *representational account*—is combined with four-dimensionalism to yield the *temporal-parts account* of temporal supervenience. This elegant account asserts and explains the theses that the facts of persistence logically supervene on facts about the spatiotemporal location of temporal parts of objects, and that the facts of temporal instantiation logically supervene on facts about the atemporal instantiation of properties by temporal parts of objects. I further show that the questions of temporal supervenience as well as the four-dimensionalist answers to these questions have interesting spatial and modal analogues. Despite its success in explaining the supervenience of facts of persistence and change, the temporal-parts account has objectionable consequences. I raise three problems, the most serious of which is the problem of predicational overkill. Each of these problems concerns the account's failure to capture certain ordinary temporal facts.

In Chapter 5, I develop a three-dimensionalist account of temporal supervenience—the *temporal-regions account*—and argue that the latter shares the main virtues and avoids the main drawbacks of its four-dimensionalist rival. The three-dimensionalist account asserts and explains the theses that the facts of persistence logically supervene on facts about the spatiotemporal location of objects, and that the facts of temporal instantiation logically supervene on the atemporal instantiation of properties by temporally unextended spacetime regions occupied by objects. I point out structural similarities of the temporal-regions account and the temporal-parts account, and show that the temporal-regions account avoids the problems that threaten the temporal-parts account. The remainder of the chapter deals with various consequences and apparent difficulties of three-dimensionalist supervenience.

1

Temporal Supervenience

The subject of this book is the problem of temporal supervenience. The aim of this chapter is to state the problem and to lay the foundations for its discussion. The problem of temporal supervenience links the project of exploring the temporal dimension of language with the project of exploring the temporal dimension of reality. I shall begin by characterizing these two projects individually and then show that they are part of, and closely linked by, the project of explaining temporal supervenience.

1.1 TEMPORAL LANGUAGE

Ordinary language is related to time by being temporally modified. When we say that something is the case, we also indicate at what time it is the case. To explore the temporal dimension of language is, therefore, to explain how temporal modification works.

The problem of temporal predication

Let us start with temporally unmodified predications, or atemporal predications, with the surface form '*a* is F', such as 'Zoe is happy', which contain a predicate 'is F' that is made up of a tenseless copula 'is' and an adjective 'F'. Temporally modified predications, or temporal predications, may be formed by temporally modifying atemporal predications of this sort. At least two kinds of temporal modification are relevant: modification by tense and modification by temporal adverbials. Consider the following examples:

(1) Zoe was happy.
(2) Zoe was happy yesterday.

The first sentence contains a predicate in the past tense. The second sentence contains, in addition to a tensed predicate, the temporal adverbial 'yesterday'. Another kind of temporal predication may be formed by temporally modifying atemporal predications with the surface form '*a* Fs', such as 'Zoe dances', which contain a verb, yielding, for example:

(3) Zoe danced yesterday.

For reasons of simplicity, I shall be concerned primarily with temporal predications such as (1) and (2). According to a widely accepted treatment of temporal predications such as (3), these are existentially quantified sentences involving a predicate with a bound event variable. (3) asserts that there exists some event that is a dance by Zoe and that occurred yesterday. The linguistic evidence for and details of this treatment as well as the metaphysical nature of events and their relation to time are issues that lie beyond the scope of this book.[1] The task of explaining how temporal modification works will thus be restricted to the task of specifying the logical form and semantics of such temporal predications as (1) and (2), as well as certain special cases, such as 'Zoe existed yesterday', that do not admit an analysis in terms of events, and are best assimilated to (1) and (2). I will refer to this task as the *problem of temporal predication.*

To specify the logical form of an ordinary, English temporal predication is to associate the English sentence with a sentence of a formal language with the purpose of elucidating the structure of temporal modification and of allowing for a clear semantic treatment. The semantics of a temporal predication will then be the semantics of its associated sentence in the formal language. I will assume for present purposes that the semantics of natural language takes the form of a T-theory.[2] A T-theory yields theorems of the form:

(T) s is true $\equiv p$

where s is an expression in the *object language*—the language that is under investigation—and p is an expression in the *metalanguage*—the language in which the investigation is conducted. The semantics of the simple temporally unmodified predication '*a* is F', which has the logical form 'F*a*', is given by the following theorem:

(T_0) 'F*a*' is true $\equiv$ F*a*

[1] For a survey of the terrain and references, see Pianesi and Varzi (2000).

[2] See Davidson (1967). The question what form a semantic theory should take is of marginal relevance for the discussion of temporal modification to follow, since none of my considerations will depend on specific features of such a theory. The discussion will be framed by truth-conditional semantics, but may equally be framed by a different semantic theory.

The right-hand side of the theorem specifies the literal truth conditions of the sentence on the left-hand side; and the truth conditions of a sentence deliver the semantic content of that sentence. The problem of temporal predication is how to extend this picture to temporally modified predications. In what follows, I shall answer part of this problem, or rather strip it down to its core, by sketching two treatments of tense and corresponding treatments of temporal adverbials.

Tenserism

In order logically to represent tense in natural language, it is common for tensers to introduce sentential tense operators. I shall assume that among those tense operators are the past-tense operator 'WAS' and the future-tense operator 'WILL'. These tense operators combine with a present-tense sentence to form a complex sentence—for example, 'F*a*' combines with 'WAS' to form 'WAS[F*a*]'. The past-tense predication '*a* was F' may then be regimented as 'WAS[F*a*]'. This account of the logical grammar of tense is essentially the account adopted by tense logicians such as Arthur Prior.[3] An alternative logical grammar is to treat the tenses not as operators on sentences, but rather as predicates that take as their arguments events, states of affairs, propositions, or proposition-like objects. On such an account, the predication '*a* was F' may be regimented as 'PAST(<F*a*>)', where <F*a*> is the state of affairs of *a*'s being F or the proposition that *a* is F.[4] I shall continue the exposition of tenserism in terms of the treatment of tenses as sentential operators.

The tenser construes her basic tense operators 'WAS' and 'WILL' as semantically primitive, which is to say that these tense operators cannot be understood in other terms and are used in the metalanguage in which the truth conditions of tensed object-language sentences are stated. (Overtly complex operators might also be used in the metalanguage but might be understood in simpler terms, and hence would not be semantically primitive.) The thesis that tense operators are primitive, and hence not to be analysed in terms of temporal singular terms, corresponds to some tensers' aim to avoid commitment to temporal entities such as instants, or times.[5] Tenserist semantic clauses for sentences modified by 'WAS' and 'WILL', and for present-tense sentences

[3] See Prior (1957, 1967, 1968*c*). [4] See, e.g., Ludlow (1999: sect. 7.2).

[5] Other tensers want to admit instants and construct them from tenserist resources; see Fine's Postscript in Prior and Fine (1977). These tenserist instants differ radically from the detenserist instants to be encountered below.

unmodified by a tense operator, may be stated as follows: for all sentences *s*,

(T_t) 'WAS'$^\frown s$ is true ≡ WAS[*p*]
s is true ≡ *p*
'WILL'$^\frown s$ is true ≡ WILL[*p*]

The problem of temporal predication is to explain how temporal modification works in the case of predications of the form 'F*a*'. Letting *s* in (T_t) be 'F*a*', the tenser proposes the following semantic axioms for modification by tense:

'WAS[F*a*]' is true ≡ WAS[F*a*]
'F*a*' is true ≡ F*a*
'WILL[F*a*]' is true ≡ WILL[F*a*]

Given that simple tenses are represented as single occurrences of tense operators, it seems natural to represent more complex tenses as multiple, nested occurrences of tense operators. Thus, while the simple future in '*a* will be F' is analysed as 'WILL[F*a*]', the future perfect in '*a* will have been F' may be analysed as 'WILL[WAS[F*a*]]'.[6]

This disquotational treatment of tense operators may also be used on certain temporal adverbials. For example, '*a* was F yesterday' may be read as 'yesterday[F*a*]', and '*a* will be F tomorrow' may be read as 'tomorrow[F*a*]'. A strategy that is more economical in that it avoids having to introduce an individual semantic axiom for 'yesterday', 'two days ago', 'last month', 'last year', and so on is to analyse these adverbials in terms of metric tense operators of the form 'WAS_n', to be read as 'it was the case *n* units of time ago', and '$WILL_n$', to be read as 'it will be the case *n* units of time hence'. On this treatment, a temporal adverbial such as 'yesterday' does not refer to a time, nor is it a predicate of times.

Other temporal adverbials do not submit to an analysis in terms of metric tense operators. Consider the adverbial 'on 20 July 2004'. This is a calendar name that seems to pick out a specific time. How should a tenser who wishes to avoid all commitment to times treat adverbials of this kind? Ludlow proposes to treat such expressions as containing an implicit when-clause of which the date is a constituent. Thus, '*a* was F on 20 July 2004' becomes 'WAS[*a* is F when [. . . 20 July 2004]]'. Ludlow suggests that the elided part of the when-clause could indicate some form of conventional dating

[6] Peter Ludlow argues that the construal of complex tenses in terms of nested tense operators is inadequate. It seems, he claims, 'that some form of temporal anaphora is necessary to account for genuine cases of past perfect, future perfect, etc.' (1999: 101). Ludlow introduces a tenserist form of temporal anaphora in (1999: ch. 8).

system, such as 'standard calendar systems indicate'.[7] The details of this proposal and the question whether there are alternative proposals need not detain us here. What matters is that there is a way of avoiding apparent commitment to times across the board.

According to tenserism, not only tensed sentences have tensed truth conditions, but all sentences have tensed truth conditions. This extension of the tenserist credo is trivial if there are no tenseless sentences. A tenser may think that this is the case, holding that an apparently tenseless sentence, such as '*a* is F simpliciter', is really an abbreviation of a disjunction of tensed sentences, '*a* was F or *a* is F or *a* will be F'. A tenser who holds that the meaning of tenseless sentences is less baroque may deny that '*a* is F simpliciter' is short for a disjunction of tensed sentences, and hence allow the predication to be genuinely tenseless, while agreeing that tenseless predications have tensed truth conditions.[8] In order to state precisely the thesis that tenseless predications have tensed truth conditions, I will introduce the sentential operator 'SIMP_t' to represent the tenser's understanding of the adverb 'simpliciter':

(4) 'SIMP_t'$^\frown s$ is true $\equiv$ $\text{WAS}[p] \vee p \vee \text{WILL}[p]$

Those tensers who hold that apparently tenseless predications are abbreviations of tensed predications may strengthen this semantic clause by replacing '$\equiv$' with '$=_{df}$', thereby turning the clause into a definition of the meaning of its left-hand side. The tenser's take on tenseless sentences may now be expressed as the thesis that all (apparently) tenseless sentences are implicitly prefixed by 'SIMP_t'.

Since *all* sentences have tensed truth conditions, metalanguage sentences as well as object-language sentences have tensed truth conditions. Accordingly, there is no genuinely tenseless metalanguage sentence 's is true simpliciter'. There are only tensed ways for a sentence s to be true: s either was, is, or will be true; and for a tenser to say that s is true simpliciter is to say that s is true simpliciter$_t$, which is equivalent to saying that s was true, is true, or will be true. So all truth is transient in the tenserist framework.

Detenserism

While the tenser says that tense is semantically irreducible, the detenser says that tense is semantically reducible; the truth conditions of

[7] See Ludlow (1999: 124–6).

[8] For discussion of this issue, see Craig (2000*a*: 4–6).

tensed sentences, along with those of tenseless sentences, are given in a tenseless metalanguage—in short, all sentences have tenseless truth conditions.

The detenser's take on tenseless predications is straightforward. In order to state precisely the thesis that tenseless predications have tenseless truth conditions, I will introduce the sentential operator 'SIMP_d' to represent the detenser's understanding of the adverb 'simpliciter':

(5) 'SIMP_d'$^\frown s$ is true $\equiv$ $\text{SIMP}_d[p]$

This operator is semantically primitive; it is an operator that cannot be understood in other terms and has access to the metalanguage, an operator that indicates that the sentence it governs is genuinely tenseless. Since the metalanguage is completely tenseless, every sentence in the metalanguage may be prefixed by 'SIMP_d'. Compare the 'SIMP_d'-operator to the 'SIMP_t'-operator. The detenser eliminates all tense from her metalanguage, and therefore does not accept the 'SIMP_t'-operator; to the detenser, 'SIMP_t', and hence sentences with tensed truth conditions, are a myth. The tenser, on the other hand, tenses every metalanguage sentence, and therefore does not accept the 'SIMP_d'-operator; to the tenser, 'SIMP_d', and hence sentences with tenseless truth conditions, are a myth.

The task of specifying tenseless truth conditions for tensed predications is harder. Let us start with the logical form of tensed predications, such as '*a* was F'. The most natural detenserist logical grammar is the following. The tense in the predication '*a* was F' is logically reducible to a predication containing the temporal indexical 'now':

(6) $\exists t(t < \text{now}\ \&\ a \text{ is F at } t)$

Here '*t*' is a variable ranging over instants, or moments. At first sight, it seems that what the past tense contributes to logical form is existential quantification over instants, the clause '$t < \text{now}$' and the temporal modifier 'at *t*'. Strictly speaking, however, all that the past tense contributes to logical form is the clause '$t < \text{now}$'—that is, the past tense contributes a predicate of instants. Similarly, the present tense contributes the clause '$t = \text{now}$' and the future tense contributes the clause '$t > \text{now}$'. What '$\exists t(\ldots \text{ at } t)$' represents is not the tense of the sentence, but rather the frequency aspect of the sentence.[9] The frequency aspect of a sentence answers the question *how often?* The default reading is *at least once*, which appears in the absence

[9] See Parsons (1990: 213–14).

of other frequency adverbials to be considered below. Thus, '*a* was F' is to be read as '*a* was F *at least once*'.[10,11]

Given that the tenses are indexical predicates, and that tensed sentences are regimented as indexical sentences quantifying over instants, what are the truth conditions of tensed/indexical sentences according to the detenser? Two types of approach to giving tenseless/non-indexical truth conditions for tensed/indexical sentences have traditionally been distinguished: the old B-theory, or old tenseless theory, of time and the new B-theory, or new tenseless theory, of time (both of which would more appropriately be called theories of tense).[12] The old B-theory says that tensed/indexical sentences are tenseless/non-indexical sentences in disguise, and hence that tensed sentences have tenseless truth conditions. The new B-theory denies that tensed/indexical sentences are tenseless/non-indexical sentences in disguise, but agrees that tensed sentences have tenseless truth conditions. I will briefly contrast three semantic accounts of '*a* is F now' and reject the first two in favour of the third. The first two are instances of the old B-theory, whereas the third is an instance of the new B-theory.

The first semantic account says that a tensed/indexical sentence is a disguised date sentence. The temporal indexical 'now' picks out different times in different contexts of utterance. When uttered at t_1, 'now' refers to t_1, when uttered at t_2, 'now' refers to t_2. So much is common ground among friends of the old and the new B-theory. Views diverge on the issue of how 'now' manages to refer to different times. According to the date

[10] One could insert an intermediate step between '*a* was F' and '$\exists t(t <$ now & a is F at $t)$' using the predicate 'is past'. So '*a* was F' is first analysed as:

(∗) $\exists t(t$ is past & a F at $t)$

Defining 't is past' as '$t <$ now', along with defining 't is future' as '$t >$ now' and 't is present' as '$t =$ now', turns (∗) into '$\exists t(t <$ now & a is F at $t)$'. As I see it, the predicates 'is past', 'is present', and 'is future' apply only to times. It is common to extend their application to events and perhaps states of affairs as well: the event of the invention of the computer is past and the state of affairs of Suzie's being a programmer is present. I find these applications misguided because natural language contains no such constructions. Events are not past; events occur in the past, the past being the set of times that are past. Likewise, states of affairs are not present; states of affairs obtain in the present, the present being the time that is present. (Where natural language allows us to apply the predicate 'is present' to things other than times is in constructions such as 'The author is present', by which we mean something like 'The author is here'.)

[11] For the standard extension of this logical treatment of simple tenses, such as '*a* will be F', to more complex tenses, such as '*a* will have been F', see Reichenbach (1947). For developments of the Reichenbachian picture, see, e.g., Hornstein (1990) and Giorgi and Pianesi (1997). For further references, see Ludlow (1999: 78).

[12] The distinction between the old and the new B-theory is discussed in Mellor (1981, 1998), Smith (1993), Oaklander and Smith (1994), Ludlow (1999), and Craig (2000*a*, *b*).

variant of the old B-theory, 'now' is ambiguous; when uttered at t_1, 'now' means the same as 't_1', and when uttered at t_2, 'now' means the same as 't_2', where 't_1' and 't_2' are schematic variables to be replaced by date adverbials, such as '26 July 2004, at 3.00 p.m.'. Moreover, when uttered at t_1, '*a* is F now' expresses the same proposition as '*a* is F at t_1', and when uttered at t_2, '*a* is F now' expresses the same proposition as '*a* is F at t_2'. The proposition expressed by an utterance of a sentence, or what is said by a sentence as uttered on a particular occasion, encodes the truth conditions of the utterance. Thus, an utterance at *t*—in short, an utterance u_t—of '*a* is F now' has the following non-indexical truth conditions:

(T_{d_1}) An utterance u_t of '*a* is F now' is true iff *a* is F at *t*.

Given the detenser's logical analysis of the tenses, (T_{d_1}) determines that an utterance u_t of '*a* was F' is true iff *a* is F at a time earlier than *t*, and an utterance u_t of '*a* is (presently) F' is true iff *a* is F at *t*, and an utterance u_t of '*a* will be F' is true ≡ *a* is F at a time later than *t*.[13]

The date variant of the B-theory faces the problem of essential indexicality. Suppose that I believe that

(7) The film starts at 8.30 p.m.

I then look at my watch and discover that 8.30 p.m. is now. I subsequently shout:

(8) The film starts now.

Here we have two utterances that do not seem to express the same semantic knowledge. I remain calm knowing that (7), but I start running as I come to realize that (8). Thus my belief that (7) is insufficient to explain why I start running. What explains my change in behaviour is the realization that (8). If the date analysis of 'now' is correct, however, an utterance of (8) at 8.30 p.m. expresses the same proposition as any utterance of (7). Thus, if the proposition expressed by an utterance of (8) at 8.30 p.m. is the object of my belief, then my belief that (8) fails to explain my change in behaviour. An utterance of an indexical sentence conveys a temporal perspective; when I say that *a* is F now, I relate *a*'s being F to my own temporal position. This relation to the speaker's temporal position is essential to explaining certain human thoughts and actions. Since the date variant of the old B-theory is unable to capture the temporal perspective induced by 'now', the date variant is explanatorily insufficient. Tensers do not face this problem. The tenser's truth conditions (T_t) treat tenses

[13] This variant of the old B-theory, or one close to it, was held by Russell (1906) and Frege (1918/1956). For detailed discussion, see Craig (2000*a*: 24–51).

disquotationally—that is, an utterance of a tensed sentence expresses a tensed proposition. Accordingly, tenses are not assimilated to dates; an utterance of 'The film is starting' and an utterance of 'The film starts at 8.30 p.m.' express different propositions. The disquotational treatment of tenses is not available to detensers, since they aim to give tenseless/non-indexical truth conditions.[14]

The second variant of the old B-theory says that a tensed/indexical sentence is a disguised token-reflexive sentence. On this variant, 'now' is not an ambiguous term with systematically varying meaning. Instead, 'now' means the same as 'the time at which this token is uttered', just as 'I' means the same as 'the person who utters this token'. Moreover, every utterance of '*a* is F now' expresses the proposition that *a* is F at the time at which this token is uttered. Thus, an utterance u_t of '*a* is F now' has the following non-indexical, token-reflexive truth conditions:

(T_{d_2}) An utterance u_t of '*a* is F now' is true iff *a* is F at the time of u_t.

Given the detenser's logical analysis of the tenses, (T_{d_2}) determines that an utterance u_t of '*a* was F' is true iff *a* is F at a time earlier than the time of u_t, and an utterance u_t of '*a* is (presently) F' is true iff *a* is F at the time of u_t, and an utterance u_t of '*a* will be F' is true iff *a* is F at a time later than the time of u_t. The token-reflexive variant avoids the problem of essential indexicality. The difference between my believing that (7) and my believing that (8) is explained by letting every utterance of (8) express the proposition that the film starts at the time at which this token is uttered. By making reference to the tokening of the sentence, the starting time of the film is related to my own temporal position, and hence the temporal perspective induced by 'now' is captured.[15]

The token-reflexive variant of the old B-theory faces the problem of tokenless truth. Consider the sentence

(9) There are no utterances now.

Application of (T_{d_2}) yields the following truth conditions of an utterance of (9):

(10) An utterance u_t of 'There are no utterances now' is true iff there are no utterances at the time of u_t.

[14] The problem of essential indexicality is forcefully presented in Perry (1977, 1979). See also Prior's case of 'Thank goodness that's over', in his (1959, 1970).

[15] This token-reflexive account is due to Reichenbach (1947). A similar account is proposed in Smart (1962, 1963). For discussion, see Craig (2000*a*: 51–64).

An utterance of (9) is false but might be true. The truth conditions in (10), however, entail that an utterance of (9) cannot be true.[16] For a detenser to capture the contingency of (9), reference to the utterance u_t must be removed from the truth conditions without removing reference to the time of u_t. Thus, (T_{d_1}) avoids the problem of tokenless truth. According to (T_{d_1}), an utterance at t of (9) is true just in case there are no utterances at t. Since it is possible for there to be no utterances at t, u_t can be true. The return to (T_{d_1}), however, is no option for the detenser, since (T_{d_1}) founders on the problem of essential indexicality. As a consequence, the old B-theorist is thrown into a dilemma: (T_{d_1}) accounts for tokenless truth but not for essential indexicality, whereas (T_{d_2}) accounts for essential indexicality but not for tokenless truth. So far, then, the detenser's case looks weaker than the tenser's. For the tenser has no more of a problem with tokenless truth than with essential indexicality, since the tensed truth conditions of tensed sentences given in (T_t) involve no reference to utterances.

The new B-theory denies that tensed/indexical sentences are disguised tenseless/non-indexical sentences, but agrees that tensed/indexical sentences have tenseless/non-indexical truth conditions. One variant of the new B-theory says that tensed/indexical sentences have the token-reflexive truth conditions stated in (T_{d_2}), but denies that tensed/indexical sentences are to be translated as token-reflexives.[17] This version of the new B-theory falls prey to the problem of tokenless truth, just as the token-reflexive variant of the old B-theory does. A more promising variant of the new B-theory is to adopt the truth conditions stated in (T_{d_1}), but to deny that tensed/indexical sentences are disguised date-sentences. This second variant may be fleshed out in terms of the now-standard semantics of indexicals most prominently held by David Kaplan and John Perry.[18] The Kaplan–Perry account distinguishes between the content of an indexical—the indexical's contribution to the proposition expressed by an utterance of a sentence in which the indexical occurs—and its 'linguistic meaning'—the rule of use that we learn when we learn a language—also known as *character* or *role*.

[16] This problem of tokenless truth is raised in Smith (1993). The problem is analogous to Castañeda's case against the token-reflexive analysis of 'I' in his (1967). Castañeda points out that an utterance of 'I am uttering nothing' is contingent. But if 'I' means the same as 'the person who utters this token', then an utterance of 'I am uttering nothing' cannot be true, since it means the same as 'The person uttering this token is uttering nothing'. The problem is also raised in Kaplan (1979, 1989) with respect to 'I am here now'.

[17] This version is proposed in Mellor (1981). For discussion, see Craig (2000*a*: 67–91).

[18] See Kaplan (1979, 1989) and Perry (1977, 1979). Castañeda (1967) must also be mentioned in this context.

The linguistic meaning of 'now' may be expressed by the token-reflexive definite description 'the time of this utterance'. This linguistic meaning together with facts about an utterance determine the time to which an utterance of 'now' refers. So far, the semantic account of 'now' resembles that of Reichenbach's token-reflexive variant of the old B-theory. The Kaplan–Perry account diverges from Reichenbach's account in that the Kaplan–Perry account leaves the linguistic meaning of 'now' out of the content; what an utterance of 'now' contributes to the proposition expressed is no more than the referent, a time. Thus, the proposition expressed by an utterance at t of 'a is F now' is the proposition expressed by any utterance of 'a is F at t'. Accordingly, the truth conditions of an utterance of 'a is F now' are those stated in (T_{d_1}).

The Kaplan–Perry account of indexicals avoids the problem of tokenless truth, since the truth conditions stated in (T_{d_1}) do not involve reference to an utterance. The problem of essential indexicality, however, still seems threatening. If my belief that the film starts now has as its object the proposition expressed by an utterance at 8.30 p.m. of 'The film starts now', then my belief that the film starts now reduces to the belief that the film starts at 8.30 p.m., which is insufficient to explain my change in behaviour. The detenser may avoid the problem of essential indexicality by construing belief and other psychological attitudes as sensitive to the token-reflexive linguistic meaning of indexical sentences.[19] One way of implementing this idea is to say that when I believe that the film starts now, the object of my belief is not the proposition expressed by an utterance of 'The film starts now'. The object of my belief is rather the token-reflexive linguistic meaning (or character or role) of the sentence 'The film starts now'. Thus, when I believe that the film starts now, I believe that the film starts at the time of this utterance. The latter belief relates the starting time of the film to my temporal position, and thereby explains why I start running. The Kaplan–Perry variant of the new B-theory thus avoids the problem of essential indexicality by giving indexical sentences token-reflexive linguistic meanings and construing belief and other attitudes as sensitive to these meanings. Moreover, it avoids the problem of tokenless truth by giving utterances of indexical sentences token-insensitive truth conditions. Henceforth, I will mean the Kaplan–Perry variant of the new B-theory when I speak of detenserism.

In addition to being tensed, ordinary language is modified by temporal adverbials. We saw with respect to tenserism that a treatment of temporal adverbials is correlated with a treatment of tense. A treatment of adverbials that is very different from the tenser's is therefore to be expected from the

[19] See Perry (1977), Lewis (1979), and Mellor (1981: ch. 5).

detenser. The detenser may distinguish between adverbials specifying an interval, such as 'yesterday' and 'between 3.00 and 4.00', and adverbials specifying an instant, such as 'yesterday at midnight' and 'on 17 February 2003 at 1.00 p.m.'. Following a suggestion by Terence Parsons,[20] I shall treat these adverbials as contributing to logical form predicates of instants, just as the detenser's tenses were above construed as contributing predicates of instants. The sentence '*a* was F yesterday', for example, receives the following reading:

(11) $\exists t(t <$ now & $t \in$ yesterday & a is F at $t)$

The clause contributed by 'yesterday' is '$t \in$ yesterday', where 'yesterday' picks out the set of all instants during yesterday. Since 'yesterday' is an interval-adverbial, it leaves open at which instant *a* is F. Consider further two cases involving instant-adverbials:

(12) *a* was F yesterday at midnight.
(13) *a* was F on 17 February 2003 at 1.00 p.m.

Parsons suggests a modular account of these complex adverbials, which splits the adverbials up into several predicates of the same instant, the intersection of which predicates effectively determines at which particular instant *a* was F:

(14) $\exists t(t <$ now & $t \in$ yesterday & $t \in$ midnight & a is F at $t)$
(15) $\exists t(t <$ now & $t \in$ 17th & $t \in$ February & $t \in$ 2003 & $t \in$ 1.00 p.m. & a is F at $t)$

Here 'midnight' stands for the set of midnights on any day, 'February' stands for the set of instants during the February of any year, '17th' stands for the set of instants on the 17th day of any month, and '1.00 p.m.' stands for the set of 1.00 p.m.-instants of any day.

In addition to interval-adverbials and instant-adverbials, there are adverbials of frequency, such as 'once', 'frequently', and 'always'. When these are absent, temporal predications involve, by default, existential quantification over instants, as was noted above in the discussion of tense. That is, the default reading of '*a* was F' is '*a* was F *at least once*'. When frequency adverbials are present, however, they can replace the default quantifiers by others.[21] Consider the following examples:

(16) *a* was always F.
(17) *a* was always F at midnight.

[20] Parsons (1990: 215–16).
[21] ibid. (1990: 210–11, 214–15).

Letting 'T' be a variable ranging over intervals, these sentences can be read as follows:

(18) $\exists T(T < \text{now} \ \& \ \forall t(t \in T \supset a \text{ is F at } t))$
(19) $\exists T(T < \text{now} \ \& \ \forall t(t \in T \ \& \ t \in \text{midnight} \supset a \text{ is F at } t))$

In both cases, the frequency adverbial 'always' replaces the default existential quantifier over instants of time by an existential quantifier over intervals and a universal quantifier over instants. So much for a rough logical treatment of various kinds of temporal adverbial on the assumption of detenserism. Since the tenser does not allow quantification over instants or intervals, the tenser has no use for this account of temporal adverbials.

For the tenser the problem of temporal predication is exhausted by the task of giving a semantic account of tense and temporal adverbials. Not so for the detenser. The detenser's treatment of tense and temporal adverbials leads some way towards answering the problem of temporal predication but does not tell the whole story. For while the detenser has logically analysed away tense and temporal adverbials, the detenser has not analysed away all temporal modification. One basic type of temporal modifier remains, namely the modifier 'at *t*', as it occurs in '*a* is F at *t*', where '*t*' is a variable ranging over instants. This is the type of temporal modification that underlies the ordinary forms of temporal modification—tense and temporal adverbials—as they occur in such predications as '*a* was F yesterday'. Therefore, it is temporal predications with the surface form '*a* is F at *t*' that pose the core problem of temporal predication for the detenser. Since I favour detenserism over tenserism, this problem will constitute my detenserist framework for further exploring the temporal dimension of language in Chapters 3–5. Since it is not possible to do the debate between tensers and detensers full justice in the present enquiry, I will rest content with motivating my preference for detenserism by means of an argument against tenserism that emerges from the thesis of temporal supervenience. This argument will be stated in Section 1.5.

1.2 TEMPORAL REALITY

From the project of exploring the relation of language and time, we move on to the project of exploring the relation of reality and time. How are the things that surround us, ordinary objects such as persons and tables, and their ordinary properties such as shapes and colours, related to time? We might further ask how events, such as parties and explosions, are related to time. But for reasons of simplicity I shall ignore events. The question about the relation of reality and time may then be split up into three questions:

(*a*) What is time?
(*b*) How are ordinary objects in time?
(*c*) How are ordinary properties instantiated in time?

One might wonder why this list does not include a fourth question: are properties in time, and if yes, how? This question, strictly and literally understood, presupposes that properties strictly and literally exist. While the existence of objects is a fairly safe bet, the existence of properties, in the strict and literal sense of 'existence', is a matter of ongoing dispute. For this reason, a neutral attitude towards the existence of properties will be assumed. The present talk of properties is intended as a convenient device, borrowed from ordinary talk, and with no commitment to the strict existence of properties. The reason, then, for not including 'are properties in time, and if yes, how?' is that it seems hard to make sense of this question without taking property-talk seriously. The case of (*c*) is different. The question of how the property of happiness is instantiated in time, which is an instance of (*c*), might as well be put as the question of how something is happy in time, which contains no reference to properties. However, when trying to generalize the question of how something is happy in time while still avoiding reference to properties, we end up with awkward formulations such as 'how is something a certain way in time?' or with metalinguistic formulations such as 'how does something satisfy a predicate in time?'. Here property-talk comes in handy by allowing the straightforward generalization (*c*).

I am primarily concerned with questions (*b*) and (*c*). The role of question (*a*) is to lay the foundation for answering questions (*b*) and (*c*); different answers to (*a*) provide different frameworks for answering (*b*) and (*c*).

Ordinary time

Question (*a*) concerns the nature of time. I shall consider two answers to (*a*), two broad conceptions of time. On the first conception, time is what I shall call *ordinary time*. This conception of time is 'ordinary' in virtue of being closely tied to ordinary temporal discourse; it is the conception of time to which we are committed in virtue of the way we speak. As a first approximation, the ordinary conception construes time as an entity that has one dimension, is distinct from three-dimensional space, and consists of past, present, and future.

There are different versions of the ordinary conception of time: *A-time* and *B-time*.[22] Since the ordinary conception is the conception of time to which our ordinary tensed discourse commits us, the metaphysical question of the nature of ordinary time is correlated with the semantic question of the status of grammatical tense—in short, the semantics of tense determines the shape of ordinary time. The tenser's thesis that all sentences have tensed truth conditions, and accordingly that truth is transient, is an expression in the formal mode of the metaphysical thesis that ordinary time is A-time. According to the A-theoretic conception, past, present, and future are irreducible. Just as grammatical tense is primitive, so the notions of past, present, and future are primitives; we cannot say what past, present, and future are in tenseless terms. In particular, past, present, and future are not meant to be explicable in terms of instants, or times, and the earlier-relation (or the later-relation) holding between times. Correspondingly, tensed facts are ultimate features of reality. Moreover, tensed facts change their temporal status. The fact that Clinton is US president was future in the distant past, present in the recent past, and is past in the present; the same fact was future, then became present, and has now become past. So reality is immersed in the irreducible flow of time, in which the future becomes the present and the present becomes the past. The combination of tenserism and A-time is commonly known as the *A-theory of time*.[23]

The version of ordinary time corresponding to detenserism is B-time. The detenser's thesis that tensed sentences have tenseless truth conditions is an expression in the formal mode of the metaphysical thesis that ordinary time is B-time. According to the B-theoretic conception, past, present, and future reduce to a system of instants, or times, that are ordered by the earlier-relation (or the later-relation). In B-time, temporal facts do not change their temporal status; no future fact can become present and then past. The facts of past, present, and future are all distinct facts holding at different times. Accordingly, the flow of time is merely apparent. Instead of being dynamic and changing, time is static and unchanging. However, while being itself unchanging, B-time allows objects to change in time (as will become apparent shortly). The combination of detenserism and B-time is commonly known as the *B-theory of time*.[24]

[22] The terms 'A-time' and 'B-time' are adapted from McTaggart's labels 'A-series' and 'B-series'; see McTaggart (1908).

[23] A-theorists include Prior (1957, 1967, 1968*c*), Gale (1968), Schlesinger (1980), Smith (1993), and Ludlow (1999).

[24] B-theorists include Russell (1906, 1915), Frege (1918/1956), Reichenbach (1947), Goodman (1951), Quine (1960), Smart (1962), Mellor (1981, 1998) and Oaklander (1991). For surveys of the A-theory and the B-theory, see Oaklander and Smith (1994) and Craig (2000*a*, *b*).

Closely related to the question of the shape of ordinary time is the question of the ontology of ordinary time. Ontology raises issues of what there is. Temporal ontology raises the issue of whether there is a past and a future. According to *eternalism*, past, present, and future are equally real. Just as distant places are no less real for being spatially distant, distant times are no less real for being temporally distant. Things existing at past and future times, such as dinosaurs and androids, are no less real than the things that exist now. From this perspective, our temporal vantagepoint, the impression that the present is special, is purely subjective and reflects our limited access to a temporally extended reality.[25] Eternalism's most prominent rival is *presentism*, according to which only present things are real. In the presentist picture, our temporal vantagepoint, the present, is an objective feature of the world.[26] Eternalism naturally appears in the company of B-time and presentism naturally appears in the company of A-time.[27] I adopt the combination of B-time and eternalism. In Section 1.5, I will give an argument against tenserism/A-time. The standard argument from relativity against presentism will be sketched in Section 2.4. For the rest of this section, I will speak of ordinary time while remaining neutral on its shape. For ease of exposition, I will help myself to B-theoretic talk of times.[28]

Spacetime

Ordinary time may be distinguished from another conception of time, according to which there is no one-dimensional time distinct from a three-dimensional space, but rather only a four-dimensional *spacetime* of which time is merely an aspect, in the sense that certain temporal relations

[25] Defenders of eternalism include Russell (1915), Goodman (1951: ch. XI), Quine (1960: sect. 36), Smart (1962), Mellor (1981, 1998), Butterfield (1984*b*), Sider (2001), and Rea (2003).

[26] On presentism, see Prior (1968*a*, *b*, 1970), Adams (1986), Merricks (1994), Bigelow (1996), Hinchliff (1996), Zimmerman (1998*b*), Ludlow (1999), Sider (1999, 2001), Crisp (2003, 2004*a*), and Markosian (2004).

[27] However, eternalism is combined with A-time by Smith (1993). And presentism, in its spatiotemporal variant, may be combined with B-time (see Sect. 4.2).

[28] The positions on tense and time listed here are not exhaustive. There is an intermediate position between detenserism and tenserism, and one between eternalism and presentism. As regards the status of tense, Michael Tooley recently proposed to follow the detenser in reducing tense, but to follow the tenser in letting tenseless propositions change in truth value. Instead of saying, however, that certain propositions were true and are now false, Tooley says that certain propositions are true *as of* one time but false *as of* another; see Tooley (1997: esp. pt. II). As regards the ontology of time, one might hold the intermediate position between eternalism and presentism, known as the theory of the growing block universe, according to which only the past and the present but not the future are real; see Broad (1923: ch. II) and Tooley (1997).

are defined on this spacetime. This revolution in the conception of space and time as we know it goes back to Hermann Minkowski, who showed that the principles of Einstein's Special Relativity follow at once from this new spacetime, and hence that the experimental results that confirm these principles also support Minkowski's declaration that, 'from now on, space for itself and time for itself should completely reduce to shadows, and only a sort of union of both ought to retain autonomy'.[29] Thus, unlike the ordinary conception, the spacetime conception is not tied to ordinary temporal thought and talk. It is rather the scientific conception of time that emerges in modern physics. Subsequent to the discovery of spacetime in the context of relativistic physics, it became clear that spacetime concepts could also provide deep insights into pre-relativistic Newtonian physics, which is why we now distinguish between different forms of classical and relativistic spacetimes.[30] These are the rough circumstances under which the spacetime conception entered modern physics. One reason why the spacetime conception enters the present discussion is that, on the assumption of some form of spacetime in response to question (*a*), questions (*b*) and (*c*) turn into serious metaphysical problems. Before we can explain what exactly these problems are, we need a simple spacetime to work with.

The spacetime conception, like the ordinary conception of time, comes in different versions. For the most part of this book, I shall work with an extremely simple pre-relativistic conception of spacetime. Excursions will be made, however, to Minkowski spacetime, the spacetime appropriate to Special Relativity. Pre-relativistic spacetime may be based on a four-dimensional manifold of primitive spacetime points, which are here understood according to substantivalism.[31] I shall assume that two temporal relations are defined on this spacetime, a relation of simultaneity and a relation of directionality. Simultaneity is here taken to be an invariant relation: two points that are simultaneous for one observer are simultaneous for any other observer. For any given point there is a well-defined set of points that are simultaneous with that point. This set of points is a three-dimensional

[29] Minkowski (1909: 104). [30] Cf. Earman (1989).

[31] Substantivalism and relationism are competing ontologies of spacetime. The substantivalist takes the physicist's talk of spacetime at face value: the points that make up spacetime are genuine entities. The relationist, on the other hand, rejects the genuine existence of spacetime points and reduces all talk of spacetime to talk of spatiotemporal relations between entities of a more safe and sane ontology. Among the two views, substantivalism ranks as orthodoxy. I shall presuppose substantivalism without argument, since the dispute between substantivalists and relationists is too large for the present discussion adequately to accommodate. Standard texts on this dispute are Sklar (1974) and Earman (1989).

hyperplane of (absolute) simultaneity. Since simultaneity is also assumed to be an equivalence relation—that is, a symmetric, reflexive, and transitive relation among spacetime points—the family of hyperplanes of simultaneity is a partition of pre-relativistic spacetime. In addition to being ordered by the simultaneity-relation, spacetime points stand in the earlier-relation (or the later-relation) to each other, which gives the spacetime a temporal direction. The earlier-relation, like simultaneity, is intrinsic to the geometry of the spacetime and independent of the states of motions of observers. This sparsely structured spacetime is all we need for present purposes. What I will say about this simple spacetime regarding its relation to ordinary time holds for more complex spacetimes as well.

The ordinary conception of time and the spacetime conception are conceptually independent; they are based on distinct conceptual schemes, distinct systems of interlinked temporal concepts. The concept of a time, or instant, is a core concept in the scheme associated with the ordinary conception, whereas the concept of a spacetime point is a core concept in the scheme associated with the spacetime conception. Moreover, the concept of a time, as well as the concept of a place, is confined to the ordinary conceptual scheme, and hence does not occur in the spacetime scheme; *vice versa* for the concept of a spacetime point. Since the ordinary conception and the spacetime conception rest on distinct conceptual schemes, they provide distinct models of the temporal dimension of reality, as opposed to different parts of the same model.

Distinguishing conceptions of time does of course not rule out that these conceptions are related by bridge principles linking ordinary temporal concepts with spatiotemporal concepts. For example, a detenser who believes in times may say that what is conceived of as a time from the ordinary temporal perspective is conceived of as a hyperplane of simultaneity, a maximal set of simultaneous spacetime points, from the spacetime perspective. But to say that a time is a hyperplane is not to say that the concept of a time is the concept of a hyperplane, just as to say that water is H_2O and pain is c-fibre excitation is not to say that the concept of water is the concept of H_2O and the concept of pain is the concept of c-fibre excitation. The bridge principle linking times with hyperplanes supports frequently encountered talk of times in the context of spacetime. Whether the conceptual resources of the two models are in fact mixed, however, is irrelevant. What matters is that the conceptual resources of each conceptual scheme are sufficient to yield a model of time.

Having sketched different answers to question (*a*), 'What is time?', let us turn to question (*b*), 'How are ordinary objects in time?', and question (*c*), 'How are ordinary properties instantiated in time?' Presupposing the ordinary conception of time, (*b*) and (*c*) have straightforward and uncontroversial answers. The answer to (*b*) is simply that ordinary objects exist at various times, or, adopting a now common term, that ordinary objects *persist* through time. The answer to (*c*) is simply that ordinary properties are instantiated, or exemplified, by ordinary objects at various times. If an ordinary object instantiates incompatible properties at different times, then the object *changes* through time. Notice that this intuitive account of change allows objects to change through B-time, which is itself static and unchanging. That questions (*b*) and (*c*) have uncontroversial answers at the level of ordinary time does not mean that facts of temporal existence and facts of temporal instantiation are metaphysically unproblematic. The problem they raise will appear when we turn to temporal supervenience.

Now consider again questions (*b*) and (*c*), this time with the spacetime conception, as opposed to the ordinary conception, in the background. Part of the answer to question (*b*) is that ordinary objects are located in, or occupy, spacetime. So much I assume to be uncontroversial. What is far from clear is *how* objects are located in, or occupy, spacetime. Within the spacetime conception, question (*b*) thus turns into a serious metaphysical problem—the problem of *spatiotemporal location*. In Chapter 2, I will develop the two main accounts of spatiotemporal location: *three-dimensionalism* and *four-dimensionalism*.

What holds for question (*b*), holds for question (*c*): a significant metaphysical problem appears only in the context of the spacetime conception. Within the ordinary conception of time, the answer to (*c*) is simply that properties are instantiated by objects at various times. Assuming the spacetime conception, however, this answer will not do. If a swarm of tiny billiard balls bouncing around in a box is a model of the behaviour of a gas, then the salient features of a gas can be exclusively described in terms of tiny billiard balls in a box. Similarly, if a four-dimensional manifold of spacetime points linked by certain temporal relations is a model of time, then the way in which a property is instantiated in time can be characterized in spacetime terms, without recourse to ordinary temporal concepts. Question (*c*) thus turns into a further metaphysical problem: which component of occupied spacetime has ordinary properties? Since property instantiation is meant to be explained purely in spacetime terms, saying that an occupant of spacetime has a property at a time does not amount to an account of spatiotemporal instantiation. The concept of a time as distinct from a place is confined to the ordinary conceptual scheme. Spacetime contains only points and regions. As a consequence, having a property in spacetime is unmodified by a time. This is the problem of *spatiotemporal instantiation*. A four-dimensionalist account of spatiotemporal instantiation, the *temporal-parts account*, will be discussed in Chapter 4, and a three-dimensionalist account of spatiotemporal instantiation, the *temporal-regions account*, will be discussed in Chapter 5. Answering the problem of spatiotemporal location and the problem of spatiotemporal instantiation constitutes a fundamental task involved in exploring the temporal dimension of reality.

After having characterized ordinary time and spacetime, as well as the metaphysical problems of spatiotemporal location and instantiation, we must turn to the relationship between ordinary time and spacetime. We must further ask how the temporal dimension of reality is related to the temporal dimension of language. These issues are addressed in the following section.

1.3 TEMPORAL SUPERVENIENCE

We distinguished between two projects: the project of exploring the temporal dimension of language and the project of exploring the temporal dimension of reality. These two projects can be pursued separately. It is possible to explore the temporal dimension of reality independently of the temporal dimension of language. More specifically, it is possible to discuss the problems of spatiotemporal location and instantiation independently of the problem of temporal predication. What is the reason, then, for pursuing these projects together, for exploring the temporal dimension of language in the same context as the temporal dimension of reality? The reason is that the two projects are part of, and closely linked by, a third project, to which I shall now turn.

The key to the link between the problem of temporal predication and the problems of spatiotemporal location and instantiation lies with the relationship between ordinary time and spacetime. So how is what goes on in ordinary time related to what goes on in spacetime? The ordinary conception is a description of the temporal dimension of reality in terms of times. The spacetime conception, on the other hand, is a description of the temporal dimension of reality in terms of spacetime points and regions. We may distinguish four different views of the relationship between these conceptions and the world. These views agree that the spacetime conception describes reality correctly but differ on the status of the ordinary conception.

The *optimist* says that the ordinary conception is distinct from the spacetime conception in virtue of being based on a distinct conceptual scheme. Moreover, the ordinary conception is true or, if nothing in reality plays the role of ordinary time perfectly, close to a true conception.[32] Further, according to the optimist, the ordinary conception, or a true

[32] In Sect. 1.6 I shall consider the issue of reality's imperfect fit of the ordinary conception with respect to relativity.

variant of it, and the spacetime conception are intimately linked by logical supervenience; once all spacetime facts are fixed, there is no room for the ordinary temporal facts to vary independently (I shall clarify this link below). The *pessimist* agrees with the optimist that the ordinary conception and the spacetime conception are distinct conceptions, and that, while the spacetime conception is true, the ordinary conception is true or close to a true conception. However, the pessimist denies that the two conceptions are linked by logical supervenience; even if all spacetime facts are fixed, the ordinary temporal facts may still vary independently. The *eliminativist* says that, while the spacetime conception is true, the ordinary conception is false and far removed from any true conception. To claim that nothing in reality answers to the ordinary concept of a time would be an eliminativist move. Finally, the *reductionist* says that, while the spacetime conception and the ordinary conception are true, they are really just one conception, because the ordinary conception is reducible to the spacetime conception. To reduce the ordinary conception to the spacetime conception is to define the meanings of ordinary temporal terms in spacetime terms. Thus, to define the noun 'time', or 'instant', as meaning the same as 'hyperplane' would be a reductionist move, whereas to claim that the concept of a time and the concept of a hyperplane are distinct concepts picking out the same things in the world would not be reductionist in the sense intended here.

The optimist and the pessimist say that the ordinary conception and the spacetime conception are distinct yet compatible conceptions. They are compatible in the sense that their ontological claims do not clash; the things that satisfy the ordinary concepts and the things that satisfy the spacetime concepts may coexist. It may even be the case that certain things fall under both ordinary concepts and spacetime concepts. This compatibility stands in no conflict with the fact that there are no times according to the spacetime conception, and the fact that there are no spacetime points according to the ordinary conception. The disagreement recorded in these statements is not ontological, but conceptual; it reflects the fact that the two conceptions are based on distinct conceptual schemes. The first statement is to be understood as saying that the spacetime conception does not employ the concept of a time; and the second statement is to be understood as saying that the ordinary conception does not employ the concept of a spacetime point. The eliminativist may deny compatibility, claiming that the things postulated by the ordinary conception could not coexist with the things postulated by the spacetime conception. The reductionist denies conceptual distinctness.

I am an optimist and find the positions of the pessimist, the eliminativist, and the reductionist unattractive. According to the pessimist, ordinary time and spacetime vary independently. The extreme result of this position is that reality has two different kinds of temporal dimension; there are different, independent kinds of time. The other two opponents of optimism are radical in a different sense. According to the eliminativist and the reductionist, our ordinary temporal conception of the world is deeply misleading. In the case of the eliminativist, this is so because the ordinary conception is false and far removed from any true conception. As a consequence, the ordinary conception cannot be taken seriously at all. In the case of the reductionist, this is so because our naive conception of time, which distinguishes times from places, is the highly developed conception of spacetime in disguise. The reductionist violently assimilates what appear to be fundamentally different conceptual schemes. The reductionist says that, when the physicists came up with spacetime theory, they did not discover a different picture of the world; they rather uncovered our own picture. According to them, we have been speaking the language of physics all along without noticing. That our common way of thinking is misleading to this extent is certainly an option. But I believe that it should not be the point of departure. Optimism should be the point of departure instead—the attitude that the ordinary conception of time is worth taking seriously, and that it bears a close relationship to the spacetime conception that is worth explaining. Pessimism, eliminativism, or reductionism should be the last exit, the exit to take only when it is clear that optimism leads nowhere. It seems, in other words, that pessimism, eliminativism, and reductionism require philosophical argument, whereas optimism does not.

As an optimist, I find it overwhelmingly plausible that all facts about ordinary time are fully determined by facts about spacetime; once all the facts about spacetime are fixed, there is no room for the facts about ordinary time to vary. The facts about times are fully determined by the facts about spacetime points and regions; the facts about temporal existence and persistence are fully determined by the facts about spatiotemporal location; and the facts about temporal instantiation are fully determined by the facts about spatiotemporal instantiation. This relation between facts about ordinary time and facts about spacetime can be made more precise by means of the notion of *supervenience.*

Supervenience

Supervenience concerns facts. A fact, as I use the term, is a truth. 'It is a fact that *p*' means the same as 'It is true that *p*', which is equivalent with '*p*'. Talk of facts is thus shorthand for talk of true propositions. True propositions are sometimes distinguished from 'bits of reality', or states of affairs, that make propositions true. For present purposes, no such truthmakers are required.

'Bits of reality', worldly facts, are here viewed as mere shadows of linguistic facts.

A rough definition of supervenience may be given as follows:

> α-facts *supervene* on β-facts if any two possible situations that are indiscernible with respect to their β-facts are indiscernible with respect to their α-facts.

Different notions of supervenience may be obtained by clarifying this definition in various ways. I shall briefly review the notions of *local* versus *global* supervenience and of *logical* versus *natural* supervenience.[33]

Global supervenience differs from local supervenience in respect of how the 'situations' in the above definition are construed. We get local supervenience by construing these situations as individuals, and we get global supervenience by construing the situations as entire possible worlds, or universes. Thus, α-facts supervene *locally* on β-facts if any two individuals, in any two possible worlds, that are indiscernible with respect to the β-facts about them are indiscernible with respect to the α-facts about them. That is, the β-facts about an individual determine the α-facts about that individual. α-facts supervene *globally* on β-facts if any two possible worlds that are indiscernible with respect to their β-facts are indiscernible with respect to their α-facts. That is, the β-facts about an entire world determine the α-facts about that world. Facts concerning the shape of a thing, for example, supervene locally on physical facts about the same thing: any two things with the same physical properties must have the same shape. Local supervenience implies global supervenience, but not *vice versa*. For example, it seems right that biological facts supervene globally on physical facts without supervening locally: any two physically indiscernible worlds are also biologically indiscernible, but two physically indiscernible organisms can be biologically different, because of differences in their respective environments. Since the distinction between local and global supervenience does not matter much to the present discussion, I shall henceforth ignore local supervenience and mean global supervenience when speaking of supervenience.

Logical supervenience differs from natural supervenience in respect of how the notion of possibility is construed. α-facts supervene *logically* on β-facts if any two logically possible worlds that are indiscernible with respect to their β-facts are indiscernible with respect to their α-facts. In this case we can say that the β-facts *entail* the α-facts, where one fact entails another if it is logically impossible for the first to hold without the second to hold as

[33] My review stays close to Chalmers (1996: 32–8).

well.[34] Logical possibility is possibility in the broadest sense, unconstrained by the laws of nature. Any world is logically possible as long as it is not contradictory. Biological facts, for example, supervene logically on physical facts. Once the physical facts are fixed, there is no logical room for the biological facts to vary independently.

Finally, α-facts supervene *naturally* on β-facts if any two naturally possible worlds that are indiscernible with respect to their β-facts are indiscernible with respect to their α-facts. A naturally possible world, sometimes also called a nomologically possible world, is a way our world could be without violating any of our laws of nature. Something is naturally possible if it could come up in our world, given the right conditions. Natural possibility is a stronger constraint than logical possibility. Any naturally possible world is also logically possible, and hence logical supervenience implies natural supervenience. But many logically possible worlds are not naturally possible. A universe without gravity, for example, is conceivable but violates the laws of nature of our world.[35]

Three forms of temporal supervenience

The intuition to be clarified is that all facts about ordinary time are fully determined by facts about spacetime; once all facts about spacetime are fixed, there is no room for the facts about ordinary time to vary independently. The notion of supervenience appropriate to capture this intuition is logical supervenience: all facts about ordinary time logically supervene on facts about spacetime. This is the general thesis of *temporal supervenience*. Note that talk of facts about ordinary time and talk of facts about spacetime are licensed by the construal of facts as true propositions. This construal further allows saying that the facts about ordinary time make up the ordinary conception of time, whereas the facts about spacetime make up the spacetime conception, and that, although different kinds of fact are involved, the two conceptions describe the same portion of reality.

Three instances of temporal supervenience are relevant in the present context:

(TS1) Facts about times logically supervene on facts about points and regions of spacetime.

[34] It would be odd to allow facts to entail each other if facts were states of affairs. Entailment is a logical relation that holds between propositions. Since my facts are true propositions, my facts can entail each other.

[35] For discussion of different notions of supervenience, see Kim (1984), Teller (1984), and Chalmers (1996: ch. 2).

(TS2) Facts about temporal existence and persistence logically supervene on facts about spatiotemporal location.

(TS3) Facts about temporal instantiation and change logically supervene on facts about spatiotemporal instantiation.

A fact about times is, for example, the fact that t_1 is earlier than t_2. This relation between times logically supervenes on relations between spacetime points and regions. The fact that an object *a* exists at a time is a fact about temporal existence. This fact logically supervenes on the relation of *a* to spacetime. The fact that *a* instantiates the property ϕ at a time is a fact about temporal instantiation. This fact logically supervenes on the relation of *a*, ϕ, and spacetime.

It will be helpful to compare temporal supervenience with some other forms of supervenience. Many common supervenience theses concern the supervenience of a set of higher-level properties on a set of lower-level properties. There is, for instance, the thesis that biological properties supervene on physical properties, and there is the thesis that mental properties likewise supervene on physical properties. To see where temporal supervenience differs from these forms of supervenience, notice that the latter usually presuppose the ordinary conception of time: higher-level properties instantiated at a time are taken to supervene on lower-level properties instantiated at a time. Temporal supervenience, on the other hand, concerns the supervenience of higher-level temporal facts on lower-level temporal facts—that is, the supervenience of facts about ordinary time on facts about spacetime. Non-temporal supervenience is supervenience within a given conception of time, whereas temporal supervenience is supervenience of one conception of time on another. According to (TS3), for example, the fact that a given property is instantiated at a time supervenes on facts about the instantiation of the same property in spacetime. This thesis applies whether we consider the shape of a person at a time or the mass of a particle at a time. Starting with the fact that something has a property at a time, the difference between non-temporal and temporal supervenience lies in whether we focus on the property or whether we focus on the time. Of course, temporal and non-temporal supervenience can be combined in a global thesis such as 'Humean Supervenience': the whole truth about a world like ours supervenes on the spatiotemporal distribution of perfectly natural, microphysical properties.[36] Although I am sympathetic to Humean Supervenience, my intention is to discuss temporal supervenience in isolation from non-temporal supervenience. In order to keep things simple, I shall primarily consider supervenient temporal facts

[36] See Lewis (1986*b*: pp. ix–xvi; 1994).

and subvenient, or 'base', temporal facts, all of which involve only ordinary objects, such as persons and tables, and ordinary properties, such as shapes and colours.

The problem of temporal supervenience

Assuming that (TS1), (TS2), and (TS3) are true, each of these theses poses its own problem. We start with fairly clean-cut facts about ordinary time and the insight that these facts logically supervene on certain facts about spacetime. Then the question arises which facts about spacetime the ordinary temporal facts supervene on. This is the first part of the problems posed by (TS1), (TS2), and (TS3). (TS1) poses the question which facts about spacetime points and regions do facts about times, such as the fact that one time is earlier than another, supervene on. (TS2) poses the question which facts about spatiotemporal location do facts about temporal existence and persistence, such as the fact that an object exists at various times, supervene on. (TS3) poses the question which facts about spatiotemporal instantiation do facts about temporal instantiation and change, such as the fact that an object has different shapes at different times, supervene on. So one task is to specify the supervenience base of ordinary temporal facts.

But that is not all. Suppose that we have specified the supervenience base of all these ordinary temporal facts. We certainly would not want to take the supervenience of ordinary temporal facts on these facts of spacetime as a fundamental, primitive fact about the world. We also want an explanation of how these facts about spacetime determine the ordinary temporal facts.[37] What is still lacking, then, is an explanatory bridge from the lower-level spatiotemporal facts to the higher-level temporal facts—a bridge to reduce the bruteness of the supervenient phenomena. Such an explanatory bridge requires an 'analysis' of the ordinary temporal facts. I shall say more about what such an analysis should look like shortly. So the problems posed by (TS1), (TS2), and (TS3) have two components: to specify which spatiotemporal facts ordinary temporal facts supervene on and to explain how they supervene. The specification of the supervenience base is the metaphysical component, whereas the explanation of supervenience is the epistemological component.

All three forms of temporal supervenience are important, but I will discuss only (TS2) and (TS3) in depth. The problems raised by (TS2) and (TS3) I will collectively call the *problem of temporal supervenience*. They link the problems of spatiotemporal location and instantiation with the problem

[37] Cf. Kim (1984: 86) and Chalmers (1996: 44).

of temporal predication. It is clear that the metaphysical component of the problem of temporal supervenience just is the problem of spatiotemporal location and instantiation. The task of specifying the supervenience base of facts about temporal existence and persistence, and of facts about temporal instantiation and change, just is the task of specifying how ordinary objects are located in spacetime and how ordinary properties are instantiated in spacetime.

The epistemological component of the problem of temporal supervenience is less obvious. We saw that, in order to build an explanatory bridge from the lower-level spatiotemporal facts to the higher-level temporal facts, an analysis of the higher-level facts is needed—that is, an analysis of ordinary facts about temporal existence, such as that *a* existed yesterday, and of ordinary facts about temporal instantiation, such as that *a* was F yesterday. Such an analysis takes the form of a semantic account of the forms of temporal modification involved in the language in which ordinary temporal facts are couched, forms of temporal modification such as tense and temporal adverbials. There is no reason, though, to expect a semantic account of temporal modification to be sufficient to provide an explanatory link from spatiotemporal facts to ordinary temporal facts. Since the optimist, unlike the reductionist, holds that ordinary temporal concepts are distinct from spatiotemporal concepts, a semantic account of temporal modification will need supplementation by various non-semantic bridge principles in order to yield a full explanation of temporal supervenience.

Since the demand for an account temporal modification is exactly the problem of temporal predication as characterized in Section 1.1, it is now clear how the problem of temporal supervenience links the problems of spatiotemporal location and instantiation with the problem of temporal predication: the metaphysical component of the problem of temporal supervenience is the problem of spatiotemporal location and instantiation, whereas the epistemological component of the problem of temporal supervenience requires an answer to the problem of temporal predication. So temporal supervenience provides a good reason for exploring the temporal dimension of language and the temporal dimension of reality in the same context. We have to look for accounts of temporal predication and of spatiotemporal location and instantiation that are such that, when combined and aided by certain bridge principles, provide an account of temporal supervenience. This is how the problem of temporal supervenience opens a new perspective on the temporal dimension of language as well as on the temporal dimension of reality, and how it imposes a considerable constraint on an account of both these dimensions.

The problem of temporal supervenience demands an explanation of the supervenience of ordinary facts of persistence and change on spatiotemporal facts. Assuming that the spacetime conception is more precise and more illuminating than the ordinary temporal conception, the problem of temporal supervenience sends us on a quest for the metaphysical nature of persistence and change. The study of the metaphysics of persistence and change in the context of temporal supervenience diverges from the standard approach to persistence and change in two significant respects. First, in the present context an understanding of the nature of persistence and change is first and foremost sought at the level of spacetime; the facts underlying persistence and change are facts about spatiotemporal location and instantiation. On the standard approach, by contrast, an understanding of the nature of persistence and change is primarily sought at the level of ordinary time. This does not mean that the standard approach fails to reach beyond what we ordinarily think and say, merely describing our actual conception of how objects persist and change. It rather means that the facts underlying persistence and change, according to the standard approach, are still temporally modified. Secondly, in the present context the metaphysics of persistence and change is constrained by the semantics of temporal predication, and *vice versa*, whereas on the standard approach semantic problems are at best perceived as being of marginal relevance to the metaphysics of persistence and change.[38] Despite various doubts I have about the standard approach, my aim is not to replace the latter; my aim is rather to broaden the picture in order to reach new results.

1.4 THE PROBLEMS OF CHANGE

The fact that objects change, that they have incompatible properties at different times, is commonly perceived as raising a special metaphysical problem, known as the *problem of change*. The question what facts underlie change is meant to be more puzzling than the question what facts underlie persistence. What gives change this particular metaphysical edge? The problem is rarely stated precisely. Here is one of David Lewis's statements of what he calls 'the problem of temporary intrinsics':

> Ordinary things, for instance we ourselves, undeniably persist through time. As we persist, we change. And not just in extrinsic ways, as when a child was born elsewhere and I became an uncle. We also change in our own intrinsic character, in the way we ourselves are, apart from our relationships to anything else. When I sit I'm bent, when I stand I'm straight. When I change my shape, that isn't a

[38] For detailed discussions of persistence and change in the standard framework, see Hawley (2001) and Sider (2001). A characterization of the standard approach will be given in Sect. 2.3.

matter of my changing relationship to other things, or my relationship to other changing things. *I* do the changing, all by myself. Or so it seems. What happens must be possible. But how? Nothing can have the two incompatible shapes, bent and straight. How does having them at different times help?[39]

Lewis, and many others, present the problem as the question of how change is possible?[40] Intuitively, to change is to have different, incompatible properties at different times. Consider the following sample report of change (for ease of exposition, I shall take a detenserist stance and consider only temporal predications with the surface form '*a* is F at *t*'):

(20) Zoe is bent-shaped at t_1 and Zoe is straight-shaped at t_2.

Lewis asks how (20) can be true, given the fact that the following statement is contradictory:

(21) Zoe is bent-shaped and Zoe is straight-shaped.

A puzzle or paradox is a contradiction that follows from intuitively plausible premisses. (20) is intuitively plausible. Therefore, if our ordinary beliefs in the truth of such statements as (20) commit us to (21), then we have a puzzle. There does not seem to be anything about (20), however, that commits us to (21), and hence there does not seem to be anything intuitively problematic about (20). Granted, Zoe does all the changing by herself, and shapes are intrinsic to her. But how does this require her to have incompatible shapes? I cannot see how. Therefore I cannot see the puzzle.

In Section 1.2, we considered the question 'How are ordinary properties instantiated in time?' from the perspective of ordinary time and from the perspective of spacetime. At the level of ordinary time, the question has an obvious and uncontroversial answer: properties are instantiated at a time. Likewise, the facts of change, of having different properties at different times, belong to the most basic and straightforward of ordinary temporal facts. There is nothing puzzling about change at the level of ordinary time. Since ordinary property-instantiation is temporally relativized, a threat of contradiction is nowhere near. At the level of spacetime, on the other hand, the question of how ordinary properties are distributed is far from obvious. As pointed out in Section 1.2, the concept of a time is not part of the conceptual scheme on which the spacetime conception is based. To say that instantiation is relativized to times would thus fail to pass as

[39] Lewis (1988: 65).

[40] For discussion of the puzzle of change, or the problem of temporary intrinsics, see Lewis (1986*a*: 202–4; 1988), Johnston (1987), Lowe (1987; 1988; 2002: ch. 3), Haslanger (1989*a*, *b*), Heller (1992), Merricks (1994), Hinchliff (1996), Zimmerman (1998*b*), Hawley (2001: ch. 1), and Sider (2001: 4.6).

a spatiotemporal account of property instantiation. For, according to the spacetime conception, nothing has a property at a time. So the problem arises of how to account for the supervenience base of ordinary facts of change. Zoe's having incompatible shapes at different times surely cannot supervene on Zoe's having incompatible shapes simpliciter. The task of specifying the spatiotemporal supervenience base of ordinary facts of change is thus threatened by contradiction: what spatiotemporal facts entail the ordinary fact that Zoe is bent-shaped at t_1 and Zoe is straight-shaped at t_2, given that there are no times, and hence no temporal relativization, in spacetime, and given that Zoe cannot be bent-shaped and straight-shaped simpliciter. Here we have a real metaphysical problem about change. But it is not, as traditionally conceived, a problem at the level of ordinary time. It is rather a problem at the level of spacetime, a problem concerning the spatiotemporal supervenience base of change.

The metaphysical problem of change may be distinguished from another problem of change, a semantic problem that, unlike the metaphysical problem, does arise at the level of ordinary time. We can agree that (20) does not look contradictory. But is it obvious that (20) does not imply a contradiction? The answer is 'no'. The semantic problem of change is the problem of explaining why a report of change such as (20) does not imply a contradiction such as (21). What this problem demands is a specification of the logical form and truth conditions of unparsed change-reports.[41] A change-report such as (20) is a conjunction of two temporal predications with the surface form '*a* is F at *t*'. The task of specifying the logical form and truth conditions of change-reports thus boils down to the task of specifying the logical form and truth conditions of temporal predications of the form '*a* is F at *t*', from which we can derive the truth conditions of the more complex change-reports. Hence, the semantic problem of change is essentially the problem of temporal predication. More precisely, the semantic problem of change represents a constraint on the problem of temporal predication: to specify the logical form and truth conditions of '*a* is F at *t*' in such a way that a report of change does not imply a contradiction. Various answers to the metaphysical and the semantic problem of change will be discussed in Chapters 3–5.

1.5 A-TIME, B-TIME, AND SPACETIME

Time may be understood as ordinary time and as an aspect of spacetime. Four different views of the relationship between these conceptions and the

[41] Cf. Lowe (1988: 73).

world were distinguished in Section 1.2. The optimist says that the ordinary conception of time logically supervenes on the spacetime conception. The pessimist says that the ordinary conception and the spacetime conception are both true but independent conceptions. The eliminativist says that, while the spacetime conception is true, the ordinary conception is false and far removed from any true conception. Finally, the reductionist says that, while the spacetime conception and the ordinary conception are true, they are really just one conception, because the ordinary conception is reducible to the spacetime conception.

How is the distinction between A-time and B-time related to the question of the relationship between ordinary time and spacetime? Eliminativists say that the ordinary conception of time is false and cannot be taken seriously. Assuming that A-theorists and B-theorists disagree over which conception of time is the correct one, not just disagreeing over which conception of time ordinary temporal discourse commits us to, and hence agreeing that some version of the ordinary conception is correct, A-theorists and B-theorists are not eliminativists. Optimists say that the ordinary conception and the spacetime conception are true and linked by logical supervenience. Pessimists say that the ordinary conception and the spacetime conception are true but unlinked. Both views presuppose that the ordinary conception and the spacetime conception are compatible, that the two conceptions can both be true. This compatibility is also a presupposition of reductionism, since reductionists say that the ordinary conception is reducible to the spacetime conception; the ordinary person's temporal language is the physicist's spatiotemporal language in disguise. I will argue in later chapters that, if ordinary time is construed as B-time, then the compatibility of the ordinary conception and the spacetime conception can be sustained. What I will argue in this section is that, if ordinary time is construed as A-time, then the ordinary conception is incompatible with the spacetime conception. The truth of the spacetime conception will thus turn out to be a threat to the A-theory of time.

A-time is incompatible with spacetime because A-theorists are tensers, and tensers are unable to give an adequate semantic treatment of truths about spacetime. Recall from Section 1.1 how the tenser and the detenser treat tenseless sentences. The tenser holds that tenseless sentences have tensed truth conditions. The '$SIMP_t$'-operator was introduced to represent the tenser's understanding of the adverb 'simpliciter':

(4) '$SIMP_t$'$^\frown s$ is true $\equiv$ $WAS[p] \vee p \vee WILL[p]$

The tenser's take on tenseless sentences may then be expressed as the thesis that all (apparently) tenseless sentences are implicitly prefixed by '$SIMP_t$'. By contrast, the detenser holds that tenseless sentences have tenseless truth conditions. The 'SIMP_d'-operator was introduced to represent the detenser's understanding of the adverb 'simpliciter':

(5) 'SIMP_d'$^\frown s$ is true $\equiv$ $\text{SIMP}_\text{d}[p]$

This operator is a semantically primitive operator that indicates that the sentence it governs is genuinely tenseless. The detenser's take on tenseless sentences may then be expressed as the thesis that all tenseless sentences are implicitly prefixed by 'SIMP_d'. Compare the 'SIMP_d'-operator to the 'SIMP_t'-operator. The detenser eliminates all tense from her metalanguage, and therefore does not accept the 'SIMP_t'-operator; to the detenser, truth simpliciter$_\text{t}$ is a myth. The tenser, on the other hand, tenses every metalanguage sentence, and therefore does not accept the 'SIMP_d'-operator; to the tenser, truth simpliciter$_\text{d}$ is a myth.

What is the relationship between ordinary temporal modification and spatiotemporal language? As pointed out in Section 1.2, ordinary time and spacetime involve fundamentally different temporal concepts. The ordinary conception is about times as distinct from places, whereas the spacetime conception is about spacetime points. There are no times and places in spacetime, and there are no spacetime points in ordinary space and time; the concepts of a time and a place are confined to the conceptual scheme of ordinary space and time, and hence do not occur in the conceptual scheme of spacetime; *vice versa* for the concept of a spacetime point.

If the optimist is right and the ordinary conception logically supervenes on the spacetime conception, then we have every reason to expect there to be various bridge principles linking ordinary spatial and temporal concepts with spatiotemporal concepts. Recall the example of times and hyperplanes; the concepts of a time and a hyperplane are distinct but satisfied by the same things. Principles linking the ordinary conception with the spacetime conception may invite talk of ordinary temporal entities, such as times, in the context of spacetime. Such talk should be assigned no more than a heuristic function; it mixes distinct models of time, based on distinct conceptual schemes, by way of bridge principles.

What holds for times, holds for past, present, and future: just as the notion of a time is confined to the ordinary conception of time, and hence there are no times in spacetime, so the notions of past, present, and future are confined to the ordinary conception of time, and hence there are no past, present, and future in spacetime. I take it to be uncontroversial that the notions of past, present, and future presuppose the distinction between time and space. The A-theorist and the B-theorist disagree on how ordinary time is to be understood, but they agree that there is a distinction between ordinary time and ordinary space. Since the distinction between time and space disappears in the spacetime conception, there are no past, present, and future in spacetime.

That past, present, and future are confined to ordinary time may also be brought out by the intuitive connection between past, present, and future and times. Even tensers who do not want to commit themselves to the existence of times may informally understand the past as the set of times that are earlier than the present time and the future as the set of times that are later than the present time. Thus, if times have no place in spacetime, then past, present, and future do not have a place either.

Given that past, present, and future as well as times are absent from spacetime in the most plausible non-reductionist picture of the relationship between the ordinary conception of time and the spacetime conception, the theoretical roles that are designed to be played by past, present, and future and times disappear as well. One such role is that of modifying a sentence. If there are no times t in spacetime, then there are no temporal modifiers 'at t' either. So truths about spacetime cannot be modified by times. What holds for times, holds for past, present, and future: if there are no past, present, and future in spacetime, then there are no tense operators 'WAS' and 'WILL', and no implicit present tense either. Moreover, since the tenser construes tenses as primitive, no reductionist move of assimilating tenses to certain spacetime concepts is available. So truths about spacetime cannot be tensed and cannot have tensed truth conditions.[42]

[42] One might question the assertion that no spatiotemporal truth can be tensed with respect to the following exotic model of spacetime in which some truths are tensed. Suppose that spacetime is temporally extended and partitioned into hyperplanes. Suppose further that one such hyperplane is metaphysically privileged: it 'glows'. The glowing hyperplane is the present. Which hyperplane glows varies over time. And this variation is irreducibly tensed: one hyperplane is glowing, but a different one will glow. So the facts about which hyperplane glows are tensed facts, while all other spatiotemporal facts are genuinely tenseless. This is a version of the moving-spotlight view of time. (Thanks to Ted Sider for suggesting that I take the latter into account. He criticizes the view in (2001: 17–21).) The problem with this picture is that it fails to take the spacetime conception seriously. The spacetime conception is meant to be conceptually independent of the ordinary conception of time. This is to say that the temporal dimension of reality can be described exclusively in spacetime terms, without the help of tensed language. According to the present version of the moving-spotlight view, however, the spacetime conception is limited in its descriptive power and dependent on the ordinary conception, in that certain temporal facts can be described only with the help of ordinary, tensed language. Such a dependence is highly implausible. The moving-spotlight view avoids this problem only if it is stated entirely in spacetime terms and understood as saying that the instantiation of the property of glowing is spatiotemporally relativized. Presumably, the view would then be that different hyperplanes glow relative to different hyperplanes. In this clothing, the view still borders on unintelligibility, but at least takes the spacetime conception seriously, in that no spatiotemporal truth is tensed. Note, finally, that the considerations to follow are independent of the issue whether the moving-spotlight view

Suppose now that an object *a* occupies a spacetime region *R*. This is an example of a tenseless truth about spacetime. According to tenserism, all (apparently) tenseless sentences have tensed truth conditions. So the sentence '*a* occupies *R*'—'O(*a*, *R*)'—is to be prefixed by '$SIMP_t$', which indicates that the sentence has tensed truth conditions:

(22) '$SIMP_t[O(a, R)]$' is true ≡ $WAS[O(a, R)] \vee O(a, R) \vee WILL[O(a, R)]$

where the disjunct 'O(*a*, *R*)' is to be read as present tensed. As a truth about spacetime, however, '*a* occupies *R*' cannot have tensed truth conditions, since tenses have no sensible application to sentences involving spatiotemporal concepts. Tenses are confined to a particular conceptual scheme, the scheme associated with the ordinary conception of time, and therefore have no home in the fundamentally distinct scheme associated with the spacetime conception. Since there are no past, present, and future in spacetime, it cannot be the case that *a* occupied *R* or that *a* presently occupies *R* or that *a* will occupy *R*. Instead, the proposition that *a* occupies *R* must be genuinely tenseless. The tenser, however, does not countenance genuinely tenseless propositions. The tenser is thus unable to give an adequate semantics for the truths about spacetime. The detenser, on the other hand, avoids this problem by giving both truths about ordinary time and truths about spacetime genuinely tenseless truth conditions.

In order to evaluate a possible reply to this objection, let us distinguish between two types of tenserism: *very serious tenserism* and *less serious tenserism*.[43] According to the very serious tenser, all sentences have tensed truth conditions. This is the standard version of tenserism that has been presupposed in the previous discussion. According to the less serious tenser, not all sentences have tensed truth conditions; there are sentences with genuinely tenseless truth conditions. The less serious tenser, like the very serious tenser, admits primitive tense operators to her metalanguage. But the less serious tenser, unlike the very serious tenser, enriches her metalanguage by using the primitive '$SIMP_d$'-operator as well. By admitting the '$SIMP_d$'-operator, the tenser is able to form genuinely tenseless as well as tensed sentences in her metalanguage, and therefore to give tenseless as well as tensed truth conditions for certain object-language sentences. With this extended repertoire the less serious tenser is able to say that all tensed object-language sentences have tensed truth conditions, whereas some, or perhaps all, tenseless object-language sentences have tenseless truth conditions, and

is correct, since none of the spatiotemporal truths to be considered would be glowing truths, if the view were correct.

[43] For these labels, see Ludlow (2004).

hence are to be prefixed by 'SIMP$_d$'. For example, the less serious tenser reads '*a* existed' as 'WAS[E*a*]' and '*a* exists tenselessly' as 'SIMP$_d$[E*a*]' and gives the following truth conditions:

(23) 'WAS[E*a*]' is true ≡ WAS[E*a*]
(24) 'SIMP$_d$[E*a*]' is true ≡ SIMP$_d$[E*a*]

If the tenser is to be able to give a semantic treatment of truths about spacetime that respects the absence of tense from spacetime, then the tenser must become a less serious tenser. With the extended tools of less serious tenserism, truths about spacetime may be given tenseless truth conditions. The statement that object *a* occupies region *R* may then be construed as being genuinely tenseless: SIMP$_d$[O(*a*, *R*)]. This way statements about A-time and statements about spacetime are given different kinds of truth conditions in the same metalanguage. A-temporal statements have tensed truth conditions, and spatiotemporal statements have tenseless truth conditions.

Less serious tenserism seems to do a better job with spatiotemporal truths than very serious tenserism. But less serious tenserism's advantage is superficial: the A-time conception is still incompatible with the spacetime conception. In order to show this, I will first argue that a less serious tenser cannot hold that an object *a* has both a tenseless existence and a tensed existence—that is, *a*'s existing simpliciter$_d$ is incompatible with *a*'s having existed and with *a*'s existing now and with *a*'s going to exist.

When a quantifier is in the scope of 'SIMP$_d$'—'SIMP$_d$[∃*x* . . .]'—then the domain of quantification includes all things that exist simpliciter$_d$. This domain may be called the 'SIMP$_d$-domain'. If an object *a* exists simpliciter$_d$, then *a* is in the SIMP$_d$-domain. If *a* exists simpliciter$_d$ and was F, then there is something in the SIMP$_d$-domain that was F: SIMP$_d$[∃*x*(WAS[F]*x*)]. In other words, if *a* exists simpliciter$_d$ and *a* was F, then we are allowed to quantify into the scope of the past-tense operator and say that there is something simpliciter$_d$ that was F. More generally, for any tense operator or sequence of tense operators 'T', the following principle holds:

(25) SIMP$_d$[E*a*] & T[F*a*] ⊃ SIMP$_d$[∃*x*(T[F]*x*)]

Let 'T' be 'WAS'. Assuming less serious tenserism, the consequent of (25), 'SIMP$_d$[∃*x*(WAS[F]*x*)]', cannot be true. According to less serious as well as very serious tenserism, tensed sentences have tensed truth conditions. Thus, the tensed sentence '∃*x*(WAS[F]*x*)' cannot be true simpliciter$_d$. It can at most be true simpliciter$_t$. In property speak, something cannot have the property of having been F simpliciter$_d$. The most the tenser can say is that something has the property of having been F simpliciter$_t$.

It should be emphasized that (25) is neutral on which tense operator or sequence of tense operators is inserted for 'T'. Irrespective of the choice of 'T', quantification into the scope of 'T' is licensed on the assumption that *a* exists simpliciter$_d$. So it does not make a difference whether 'T' is 'WAS' or 'WAS[WAS[...]]' or 'SIMP$_t$[...]'. In each case the consequent is a tensed sentence that is true or false simpliciter$_d$. This result is unacceptable even to the less serious tenser. Since the consequent of (25) cannot be true with less serious tenserism in the background, the less serious tenser cannot hold that an object has both a tenseless existence and a tensed existence—that is, the tenser cannot hold that SIMP$_d$[E*a*] & T[E*a*].

This result is perhaps not worrying if nothing in the SIMP$_d$-domain is the subject of a tensed truth. But consider the inhabitants of A-time and of spacetime. Some objects are surely at home in both A-time and spacetime. Perhaps ordinary macro-objects inhabit spacetime along with microphysical objects. Perhaps, though, ordinary objects are confined to ordinary time or perhaps there are no macro-objects over and above arrangements of micro-objects. Still, there are microphysical particles that inhabit both ordinary time and spacetime. Those who doubt the occurrence of the particles of physics at the level of ordinary time should remember that Einstein originally stated his theory of relativity in ordinary temporal terms. Suppose, then, that *a* inhabits A-time and spacetime. At the level of A-time, tensed predications are true of *a*, and *a* has a tensed existence—T[E*a*]. At the level of spacetime, genuinely tenseless predications are true of *a*, and *a* has a tenseless existence—SIMP$_d$[E*a*]. But we just saw that *a* cannot have both a tensed existence and a genuinely tenseless existence. Hence, truths about spacetime pose a serious problem for the very serious tenser and the less serious tenser alike.[44]

A-temporal truths are irreducibly tensed. Correspondingly, A-time is a dynamic river. Spatiotemporal truths, like B-temporal truths, are genuinely tenseless. Correspondingly, spacetime, like B-time, is a static block—or less a block than a static slice (see presentism about spacetime in Section 2.4). Since the river is irreducible—A-time cannot be explained in B-terms—the

[44] The shape of spacetime, whether spacetime is a temporally extended block or a temporally unextended slice—see eternalism and presentism about spacetime in Sect. 2.4—and whether spacetime is Newtonian or Minkowskian, is irrelevant to the present argument for the incompatibility of A-time and spacetime. The argument relies on the premiss that the temporal dimension of reality can be described exclusively in spacetime terms, without recourse to the concepts of past, present, and future, and accordingly without the help of tensed language. The absence of ordinary temporal concepts from the spacetime conception is a feature of this conception irrespective of whether or not it characterizes time as extended and whether or not it characterizes time as relativistic.

river cannot coexist with the block, or the slice. I conclude that the deeply entrenched conception of time as an aspect of spacetime poses a threat to the conception of time as A-time. The A-theorist's last resort may ultimately be the position that the ordinary conception of time is true, but the spacetime conception is not.[45] This is a price for A-time that is not worth paying. Therefore, I will from now on hold that tense is semantically reducible and that ordinary time has the shape of B-time.

1.6 SUPERVENIENCE AND RELATIVITY

So far the discussion of the problem of temporal supervenience has been silent on the impact of relativity theory on ordinary time, spacetime, and temporal supervenience. I will close this chapter by factoring in Special Relativity (SR) and reformulate the problem of temporal supervenience as the *problem of relativistic supervenience.*

What is time? How are ordinary objects in time? And how are ordinary properties instantiated in time? These are the questions that shape the present discussion of the metaphysics of time. We distinguished between the ordinary conception of time and the spacetime conception as different answers to the first question. Then we saw that the other two questions have simple and obvious answers within the conception of ordinary time: ordinary objects exist at various times, they persist, and they have various properties at various times. Given that all ordinary temporal facts logically supervene on spacetime facts, the resulting problem was to specify which spacetime facts ordinary facts of persistence and temporal variation supervene on, and to explain how they supervene. This is the problem of temporal supervenience.

In order to set up the problem of relativistic supervenience, let us again start with our three questions and give answers that are appropriate to SR. First of all, relativistic time and space, like pre-relativistic time and space, have one temporal dimension and three spatial dimensions. In relativistic time, however, no temporal relation is instantiated absolutely. It is not meaningful to ask whether something that happens at t_1 is simultaneous with something that happens at t_2. Nor is it meaningful to ask whether t_1 is earlier than t_2 or whether t_1 is later than t_2. Similarly, no spatial relation is instantiated absolutely. Instead, all temporal and spatial relations are relativized to *frames of reference.* A frame of reference is defined by an inertial (unaccelerated) path of an observer, together with a specification of spatial axes. Every temporal fact that holds absolutely in pre-relativistic time

[45] This position is perhaps adopted in Smith (1993) and Craig (2001).

holds relative to some frame of reference in relativistic time. If we think of pre-relativistic time as a string of ordered times—a one-dimensional real line—then relativistic time is a multiplication of such strings, each of which is relativized to a particular frame of reference.

The ordinary conception of time is the conception inherent in our ordinary way of speaking. The optimist who takes ordinary time seriously can adopt one of two perspectives on relativity. The first perspective is to say that, since our ordinary thought and talk presupposes absolute temporal relations, and frames of reference play no role in everyday linguistic transactions, the ordinary conception is an incorrect description of our relativistic world. However, the ordinary conception is close to a true conception that is relativistically acceptable, and that employs relativistic counterparts of the ordinary conception's core concepts. The second perspective is to say that our ordinary temporal concepts belong to a conceptual scheme in which all entries are relativized to frames of reference. Thus, the ordinary conception of time is not to be replaced by a relativistic one. Instead, the ordinary conception is a part of the relativistic conception; our ordinary temporal concepts belong to a scheme, parts of which are hidden from the ordinary speaker. Accordingly, '*a* was F' is strictly true in virtue of containing a hidden relativization to a frame of reference: '*a* was F relative to *f*'. So ordinary time is a correct but incomplete description of the world; our ordinary temporal concepts are embedded in a bigger, relativistic picture.

The spacetime appropriate to SR is *Minkowski spacetime*. Recall that pre-relativistic spacetime may be based on a four-dimensional manifold of primitive spacetime points. In this spacetime, simultaneity is assumed to be an invariant notion. For any given point there is a well-defined set of points that are simultaneous with that point. This set of points is a hyperplane of simultaneity. Since simultaneity is also assumed to be an equivalence relation—that is, a symmetric, reflexive, and transitive relation among spacetime points—the set of hyperplanes of simultaneity is a partition of pre-relativistic spacetime. Minkowski spacetime also contains a four-dimensional manifold of spacetime points, but differs from pre-relativistic spacetime in that the notion of simultaneity is not an invariant notion; it is not meaningful to ask whether two spacetime points are simultaneous. Thus, Minkowski spacetime is not partitioned into ordered hyperplanes of simultaneity the way pre-relativistic spacetime is.

Although absolute simultaneity is not well defined in Minkowski spacetime, it is possible to define a *relative* notion of simultaneity by means of Einstein's light-signalling method. Imagine an observer moving on some inertial path, f. Such a path together with a specification of spatial axes defines a frame of reference. At a certain point, p_1, along path f, the observer sends out a light signal, which is reflected at a point p and intersects f at another point p_2. On Einstein's definition, point p is regarded as being simultaneous with the midpoint, m, between p_1 and p_2 on f, relative to the observer's state of motion along path f. This definition has the consequence that points other than m will be regarded as simultaneous with p by an observer who is in motion relative to the observer on path f. Simultaneity thus relativized to an inertial observer is an equivalence relation. And so each inertial frame defines a different slicing of the spacetime into hyperplanes of simultaneous points.

Given this definition of simultaneity relative to a frame of reference, one can also introduce frame-relative relations of 'earlier than' and 'later than'. A point p may be defined as being earlier than another point q relative to an inertial frame f iff p is in the absolute past of some point simultaneous with q relative to f. Point p may be defined as being later than point q relative to f iff p is in the absolute future of some point simultaneous with q relative to f. The notions of 'absolute past' and 'absolute future' are defined in the context of the light-cone structure of Minkowski spacetime. The light-cone structure is a specification, for every spacetime point p, of (i) the set of points with lightlike separation from p, (ii) the set of points with timelike separation from p, and (iii) the set of points with spacelike separation from p. First, points p and q have lightlike separation if a light signal emitted from one of the two points can reach the other. The set of points that have lightlike separation from a given point, p, is called the light-cone for p. Secondly, points p and q have timelike separation if a causal signal travelling below the speed of light can get from one of these points to the other. The set of points that are timelike separated from p can be further divided into (*a*) the ***absolute future*** of p: the set of points that can be reached from p by a causal signal travelling below the speed of light, and (*b*) the ***absolute past*** of p: the set of points from which p can be reached by a causal signal travelling below the speed of light. Thirdly, points p and q have spacelike separation if no causal signal can get from one point to the other unless it travels faster than light. Since a fundamental assumption of relativistic theories is that there are no such faster-than-light signals, spacelike separated points have no possible causal signal connecting them.

How are relativistic time and Minkowski spacetime related? An answer to this question depends on the ontological status of Minkowski spacetime. One view is to interpret Minkowski spacetime *instrumentally*. On this view, the spacetime has no ontological significance; it is not meant to tell us about fundamental constituents of the world. Instead, Minkowski spacetime is merely a geometrical representation of relativistic space and time, which alone carry ontological weight. Another view is to interpret Minkowski spacetime *realistically*. On this view, the spacetime has physical existence. Spacetime points and regions are not just mathematical metaphors; they are among the most fundamental entries in our ontological inventory. The realistic interpretation was adopted by Minkowski himself as well as by Einstein, although Einstein's original formulation of SR was metaphysically a theory of ordinary space and time. The realistic interpretation is the interpretation that I shall assume. From the point of view of the realist, Minkowski spacetime is best characterized as the metaphysical and explanatory basis of relativistic time. That Minkowski spacetime is the metaphysical basis of relativistic time may be put more precisely by saying that all facts about relativistic time logically supervene on facts about Minkowski spacetime. This is the thesis of *relativistic supervenience*.

How are ordinary objects in relativistic time? And how are ordinary properties instantiated in relativistic time? The pre-relativistic answers to these questions are that ordinary objects exist at various times and have various properties at various times. These ordinary temporal facts of persistence and variation have relativistic analogues: an ordinary object exists at various times relative to a given frame of reference, and an ordinary object has various properties at various times relative to a given frame of reference. Thus, in the context of SR, temporal predications are relativized to frames of reference: a is F at t relative to f. Reports of relativistic temporal variation then have the form: a is F at t_1 relative to f and a is not F at t_2 relative to f, where t_1 is earlier than or later than t_2 relative to f. These facts of persistence and change are familiar facts in new clothing. They are not, however, the whole relativistic story about how objects are in time and about how properties are instantiated in time. For relativistic time hosts facts that we have not encountered before. First, there are facts of *relativistic persistence*: an ordinary object can exist at a time t_1 relative to a frame f_1 and exist at a time t_2 relative to a frame f_2, where t_1 and t_2 are individuated relative to different frames and bear no frame-relative temporal relations to each other. Intuitively speaking, ordinary objects do not persist only within a given frame of reference: they also persist across reference frames; they are at home in different frames. Secondly, there are facts of *relativistic variation*: an ordinary object can have a certain property P at time t_1 relative to a frame f_1 and have an incompatible property Q at time t_2 relative to frame f_2, where again t_1 and t_2 are individuated relative to different frames and bear no frame-relative temporal relations to each other. In other words, ordinary objects do not vary in their properties only within a given frame of reference. Objects also vary in their properties across frames of reference.

An instance of relativistic variation is the phenomenon of length contraction. SR predicts that, given two objects in relative motion, each is contracted relative to the other. This reciprocity is a result of the fact that in SR there is no absolute space in which an object has a true length.

For illustration of length contraction, consider the following well-known thought experiment. A pole vaulter running at near light speed must run through a shed with doors at each end. Suppose that the shed's proper length is 10 metres and the pole's proper length is 20 metres, where the proper length of an object is the length of the object in its rest frame. Since the vaulter is running near light speed, the pole will, relative to the frame of reference of the shed, be contracted to half its length and so will fit inside the doors of the shed. In the frame of reference of the runner, on the other hand, the pole will be uncontracted and instead the shed will be contracted to the length of 5 metres, in which case the pole will not fit inside the doors of the shed. Is the pole in the shed or not? Since in SR there is no absolute frame of reference, there is no absolute truth about whether the pole is in the shed or not; it is relative to one frame, it is not relative to another. Notice that this variation in length across reference frames, unlike temporal variation in length inside a given reference frame, is not a dynamic phenomenon. Relativistic variation in length has nothing to do with the mechanical contraction of the length of an object under pressure. It is purely relativistic in nature.

Relativistic supervenience is the thesis that all facts about relativistic time logically supervene on facts about Minkowski spacetime. Accordingly, the facts of temporal persistence and variation and the facts of relativistic persistence and variation logically supervene on certain facts about Minkowski spacetime. The resulting problem is to determine which spacetime facts these facts about relativistic time supervene on, and to explain how they supervene. This is the problem of relativistic supervenience (to be discussed in Sections 2.1 and 5.4).

The thesis of temporal/relativistic supervenience is that all facts about ordinary/relativistic time logically supervene on facts about spacetime. This schematic thesis would be true if spacetime were Newtonian or if spacetime were Minkowskian, and it is true for the curved, general relativistic spacetime we in fact inhabit. I will rest content with discussing supervenience on classical spacetime and on Minkowski spacetime, since the explanation of these forms of temporal supervenience is sufficient to yield major results on the metaphysical debate about the nature of persistence and change. I think that these metaphysical results hold for general relativistic spacetime as well, but it lies beyond the scope of the present enquiry to show this.

So the forms of temporal supervenience being considered here are only the most basic forms. There are other, more complex forms. On the one hand, one might take into account a more complex framework for the supervenience base, as encountered in General Relativity. On the other hand, one might take into account more complex supervenient phenomena—for example, epistemological phenomena concerning how we experience ordinary time, such as our sense that the present is special and that time flows. In other words, temporal supervenience not only connects the semantics of time and the metaphysics of time, but it also connects the epistemology of time and the physics of time. Temporal supervenience is the thesis that these areas form a continuum, a continuum that holds many mysteries.

2

Three-Dimensionalism and Four-Dimensionalism

How are objects in time? Assuming the ordinary conception of time, this question has an obvious answer: objects persist through time. Assuming the conception of time as a 'shadow' of spacetime, the question becomes the problem of spatiotemporal location, which lacks an obvious answer: how are objects located in spacetime? In this chapter, I will provide a detailed statement of various answers to this problem. Knowing the possible forms of spatiotemporal location will be crucial for structuring the ensuing discussion of the problem of temporal supervenience (Chapters 3–5), of which the problem of spatiotemporal location is one component. The main answers to the problem of spatiotemporal location will be called *three-dimensionalism* and *four-dimensionalism*.

2.1 OBJECTS IN SPACETIME

Before we can discuss an object's mode of spatiotemporal location, we need to say a little more about spacetime itself. I shall state three-dimensionalism and four-dimensionalism on the basis of the pre-relativistic, substantivalist spacetime of Section 1.2. This spacetime contains a four-dimensional manifold of spacetime points and absolute simultaneity. The primitive spacetime points are the fundamental constituents of the manifold. In addition to spacetime points, there are spacetime regions. Regions are mereological constructions from points: any mereological sum of spacetime points counts as a spacetime region. Given spacetime regions, we can define the notions of a temporally unextended, or instantaneous, spacetime region and of a temporally extended spacetime region:

(IR) A spacetime region R is *temporally unextended*, or *instantaneous* $=_{\text{df}}$ R is entirely constituted by simultaneous spacetime points.

(ER) A spacetime region R is *temporally extended* $=_{\text{df}}$ R is a sum of instantaneous regions, each of which lies on a different hyperplane of simultaneity.

Moving on from spacetime regions to their relation to objects, we need the notion of spatiotemporal *occupation*, or *location*. Occupying a spacetime region simpliciter is the spacetime-analogue of occupying a spatial region at a time. For the purpose of defining three-dimensionalism and four-dimensionalism, I shall take the predicate 'occupies', as it occurs in 'an object occupies a spacetime region', as an undefined primitive. This is merely a methodological decision, not an ontological thesis—that is, I leave it open whether this predicate can be defined. I doubt, however, that the predicate can be defined as long as spacetime points and regions are treated as ontological primitives, as is standard in substantivalist spacetime theory.

For an object to occupy a spacetime region is for the object to fit into the region perfectly, like a hand fits into a glove. In place of a definition of occupation, I offer the following characterization *via negationis*. First, if an object a occupies a region R, and R is a proper part of R', then it does not follow that a occupies R'. For example, a leg of a table does not occupy the region occupied by the table. The leg merely occupies a subregion of the table's region. Secondly, if a occupies R, and R' is a proper part of R, then it does not follow that a occupies R'. For example, a table does not occupy the region occupied by its legs. The table merely occupies a region that has the leg's regions as proper parts. Thirdly, if a occupies R, then it does not follow that a occupies no region other than R. Thus, objects can occupy multiple regions. Fourthly, if a occupies multiple regions, then it does not follow that a occupies the mereological fusion of these regions. Thus, objects can occupy multiple temporally unextended, or instantaneous, regions without also occupying a temporally extended region (I shall return to this last condition below).[1]

I will now define three-dimensionalism and four-dimensionalism as two rival accounts of an ordinary object's mode of spatiotemporal occupation. Such an account needs to answer the following two questions. Is the occupation-relation one–one or one–many? And what kind of spacetime region does an ordinary object occupy? Three-dimensionalism and four-dimensionalism may be stated as the following theses and illustrated by Figure 1:

(3D) (i) an ordinary object occupies multiple spacetime regions, and (ii) these spacetime regions are temporally unextended, or instantaneous, and non-simultaneous.

[1] Thanks to Kris McDaniel for discussion of occupation.

(4D) (i) an ordinary object occupies a unique spacetime region, and (ii) this spacetime region is temporally extended.

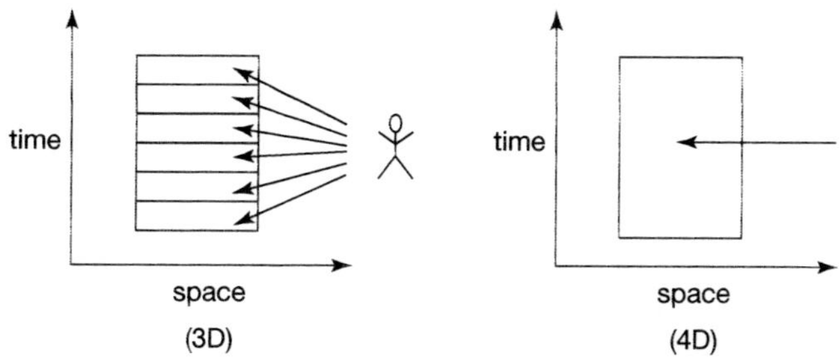

Figure 1

According to (3D), the occupation-relation is one–many, and the kind of spacetime region occupied is temporally unextended. According to (4D), the occupation-relation is one–one, and the kind of spacetime region occupied is temporally extended. In short, (3D) says that ordinary objects are not extended along the temporal dimension (but only along the spatial dimensions)—and (4D) says that ordinary objects are extended along the temporal dimension (as well as along the spatial dimensions). Moreover, since (3D) and (4D) concern only temporal, as opposed to spatial, features of spatiotemporally located objects, (3D) and (4D) may be seen as different views about how ordinary objects are in time.

Theses (3D) and (4D) are exclusive. Assuming that clauses (3D)(i) and (4D)(i) contain the same predicate 'occupies', these clauses are incompatible. To say that an object occupies both one region and many regions is just as contradictory as saying that someone loves both one woman and many women. Moreover, it is important not to lump together clauses (i) and (ii) of each thesis in such a way that (3D) becomes the thesis that an ordinary object occupies multiple instantaneous spacetime regions, and (4D) becomes the thesis that an ordinary object occupies a unique extended spacetime region. The resulting theses have a reading that renders them compatible. Compare saying that someone loves one blonde woman but many brunette women, which is not contradictory on its most natural reading. Separating clauses (i) and (ii) rules out this kind of reading of (3D) and (4D).

It has been argued that three-dimensionalism is incoherent on the ground that an object *a* cannot occupy multiple temporally unextended, or instantaneous, regions without also occupying a temporally extended region. But if *a* occupies an extended region as well as various unextended regions, then clause (3D)(ii), which says that all regions occupied are unextended, is false. In short, a three-dimensionalist object ends up being both temporally unextended and temporally extended, which is contradictory.[2] This incoherence claim is based on the premiss that an object cannot occupy multiple regions without also occupying the fusion of these regions. Here is a simple example that illustrates the falsity of this premiss. Suppose that a person travels back in time to meet her younger self, and that when the older self meets the younger self, the older self stands 10 metres away from the younger self. Here we have a situation in which one and the same person occupies distinct regions but obviously fails to occupy the sum of these regions. For the sum of the regions occupied by the older self and the younger self is a disconnected region with parts that lie 10 metres away from each other. But persons obviously cannot occupy scattered regions of this kind; no person fits perfectly into such a region. Perhaps there is some sense in which the person who occupies both regions 'fills up' the sum of these regions. But 'filling up' a region is not occupying a region, in the sense intended here. Therefore, three-dimensionalism is coherent.

Theses (3D) and (4D) are not jointly exhaustive. There are a number of conceivable mixtures. One might, for example, claim that an ordinary object occupies multiple spacetime regions, but that these regions are temporally extended.[3] Or one might claim that some regions occupied are temporally extended and others are temporally unextended. These variants will not be discussed further. Alternatively, one might claim that an ordinary object occupies a unique spacetime region, but that this region is temporally unextended. This variant will be considered below.

It must further be emphasized that (3D) and (4D) are rival accounts of how objects are in spacetime, and cannot be restated at the level of ordinary time. According to the spacetime conception, there are spacetime regions. According to the ordinary conception, there are places and times. According to the spacetime conception, objects occupy spacetime regions. According to the ordinary conception, objects exist at times and occupy places at times. To say that objects exist at different times constitutes an answer to the question how objects are in time in ordinary temporal terms. To say that objects occupy different times, on the other hand, does not receive any sanction from ordinary use, and therefore does not constitute an answer to the question how objects are in time in ordinary temporal terms. Given that objects are in ordinary time by existing at various times, there is no straightforward analogue of three-dimensionalism and four-dimensionalism

[2] See Barker and Dowe (2003) and Barker and Dowe (2005). For a reply, see McDaniel (2003).

[3] This sort of view is adopted in Hudson (2001: ch. 2).

at the level of ordinary time. All sides agree that objects exist at different unextended times: instants—this is just the persistence intuition. Moreover, all sides agree that objects persist through extended times: intervals. Thus, the ordinary conception raises no dispute about how objects are in time that mirrors the dispute between three-dimensionalists and four-dimensionalists about how objects are in spacetime.

A related point of importance is that (3D) and (4D) are stated in terms of the spatiotemporal notion of occupying a spacetime region simpliciter, and must not be interpreted in terms of the ordinary temporal notion of occupying a place at a time. For, if (3D) and (4D) were mistakenly construed as employing the temporal notion or as being explicable in terms of the temporal notion, then it would be difficult to make sense of the dispute between three-dimensionalists and four-dimensionalists. It would perhaps be plausible to understand '*a* occupies an instantaneous spacetime region' as '*a* occupies a place at a time', and hence (3D) would have a chance of making sense. But it would be hard to interpret '*a* occupies an extended spacetime region' in terms of the notion of occupying a place at a time, and hence (4D) would not even begin to make sense. Since the atemporal notion of occupation is not to be explained in terms of the temporal notion, no such obvious difficulty arises.

Theses (3D) and (4D) concern modes of spatiotemporal location in Newtonian spacetime. (3D) and (4D) are relativistically unacceptable, since they are stated in terms of the notions of a temporally unextended, or instantaneous, and a temporally extended spacetime region, which presuppose absolute simultaneity between spacetime points (see (IR) and (ER)). Relativistically acceptable formulations of three-dimensionalism and four-dimensionalism may be based on Minkowski spacetime. Minkowski spacetime differs from pre-relativistic spacetime in that the notion of simultaneity is not an invariant notion. We saw in Section 1.6, however, that it is possible to define a frame-relative notion of simultaneity. Assuming this notion, relativistic versions of the notions of a temporally unextended spacetime region and a temporally extended spacetime region may be defined in terms of frame-relative simultaneity as follows:

(IR_{SR}) A spacetime region R is *temporally unextended*, or *instantaneous*, relative to a frame of reference $f =_{df} R$ is constituted by spacetime points which are simultaneous relative to f.

(ER_{SR}) A spacetime region R is *temporally extended* relative to a frame of reference $f =_{df} R$ is a sum of regions that are instantaneous relative to f, and each of these regions lies on a different hyperplane relative to f.

With these definitions in place, (3D) and (4D) may be rendered relativistically acceptable:

($3D_{SR}$) (i) an ordinary object occupies multiple spacetime regions, and (ii) these spacetime regions are temporally unextended, or instantaneous, relative to some inertial frame of reference, and lie on different frame-relative hyperplanes.

($4D_{SR}$) (i) an ordinary object occupies a unique spacetime region, and (ii) this spacetime region is composed of regions that are temporally extended relative to some inertial frame of reference.

Clause (ii) in each thesis permits relativization to distinct frames of reference. Thus, relativistic three-dimensionalism allows an object to occupy several regions that lie on hyperplanes that are relativized to distinct frames of reference. And relativistic four-dimensionalism allows the unique region occupied by an object to be composed of regions that are temporally extended relative to distinct frames of reference.

Having stated these relativistic variants of three-dimensionalism and four-dimensionalism, it is easy to dispel a superficial motivation for the claim that SR threatens three-dimensionalism. The threat might be viewed as arising from the fact that it is common in spacetime theory, as required by SR, to represent objects by world-lines or world-tubes—that is, by timelike extended spacetime regions. Given this representation, one might be tempted simply to identify objects with world-lines or world-tubes. To make such an identification, however, would be to ignore the main issue between three-dimensionalists and four-dimensionalists: does an object occupy the whole of its world-line or world-tube, or does an object merely occupy parts of its world-line or world-tube? For an object to occupy parts, as opposed to the whole, of its world-line or world-tube, is a way for the object to be located in four-dimensional spacetime without itself being four-dimensional.[4]

In order to keep things simple, I will work with (3D) and (4D) in the discussion of temporal supervenience until I return to ($3D_{SR}$) and ($4D_{SR}$) in Section 5.4, when I take up relativistic supervenience.

2.2 TEMPORAL PARTS AND TEMPORAL COUNTERPARTS

In order to deepen our understanding of (3D) and (4D), it will be helpful to point out some extensions and consequences of the two positions. The

[4] See Rea (1998) and Balashov (1999: 322). More potential worries for three-dimensionalism arising from relativity will be discussed in Sect. 5.4.

consequences to be considered in this and the following sections are all of a taxonomical nature. They concern the relations of (3D) and (4D) to other theses and theories.

Temporal parts

(4D) says that objects are extended along the temporal dimension. Given such temporal extension, it seems natural to say that ordinary objects are divided into temporal parts along their temporal dimension, just as they are divided into spatial parts along their spatial dimensions. Since (4D) does not mention temporal parts, a theory of temporal parts must be understood as an addition to, or development of, (4D).

A *theory of temporal parts* may be stated in three theses. The first thesis, (T1), is (4D). Before the second thesis can be stated, a number of definitions are needed. Spacetime regions have smaller regions or points as parts. Among the parts of spacetime regions, we may distinguish between spatial and temporal parts. Spatial parts are not of particular interest in the present context and so may be set aside (but see Section 4.3). What, then, are temporal parts of spacetime regions? We may begin by defining instantaneous temporal parts of regions.

(ITP:R) A spacetime point or region R is an *instantaneous temporal part* of a spacetime region $R' =_{\text{df}} R$ is a maximal sum of simultaneous parts of R'.[5]

Extended temporal parts of regions may then be defined as follows:

(ETP:R) A spacetime region R is an *extended temporal part* of a spacetime region $R' =_{\text{df}} R$ is the sum of at least two instantaneous temporal parts of R'.[6]

That spacetime regions have temporal parts is not the consequence of a special thesis, but rather follows directly from the standard ontology of spacetime. A spacetime region has smaller regions or points as parts, and temporal relations are relations between spacetime points and regions. So

[5] This definition may be spelled out as follows:

(ITP:R) A spacetime point or region R is an *instantaneous temporal part* of a spacetime region $R' =_{\text{df}}$ (i) R is a part of R', (ii) the spacetime points in R are all simultaneous, and (iii) no spacetime point that is in R' but not in R is simultaneous with all spacetime points in R.

[6] Since an instantaneous temporal part of R' is a *maximal* sum of simultaneous parts of R', it follows that an extended temporal part of R' is a sum of *non-simultaneous* instantaneous temporal parts of R'.

a region R has parts that are maximal sums of simultaneous subregions of R. Hence, a spacetime region has instantaneous temporal parts according to (ITP:R). Moreover, every set of instantaneous temporal parts of a region has a fusion. As a result, a spacetime region has extended temporal parts according to (ETP:R). Assuming definitions (ITP:R) and (ETP:R), we may define instantaneous and extended temporal parts of objects, as opposed to regions, as follows:

(ITP:O) An object x is an *instantaneous temporal part* of an object $y =_{df}$ (i) x is a part of y, (ii) y occupies a spacetime region R, (iii) x occupies an instantaneous temporal part of R, and (iv) x does not occupy any other spacetime region.

(ETP:O) An object x is an *extended temporal part* of an object $y =_{df}$ (i) x is a part of y, (ii) y occupies a spacetime region R, (iii) x occupies an extended temporal part of R, and (iv) x does not occupy any other spacetime region.[7]

We now come to the second thesis of the theory of temporal parts, which correlates the temporal parts of spacetime regions with parts of objects occupying regions. The thesis may be called the *doctrine of arbitrary temporal parts*:

(T2) If an ordinary object x occupies a spacetime region R, then for every instantaneous and extended temporal part R' of R there is a part x' of x such that (i) x' occupies R', and (ii) x' does not occupy any other spacetime region.

Given definitions (ITP:O) and (ETP:O), it follows from (T2) that, if x occupies region R, then x has instantaneous and extended temporal parts that occupy corresponding instantaneous and extended temporal parts of R.[8] Thus, it follows from (T1)/(4D) and (T2) that ordinary objects have instantaneous and extended temporal parts.

Given that ordinary objects have instantaneous and extended temporal parts, an ordinary object may be characterized as a sum of temporal parts

[7] There is a structural similarity between these two definitions and definition (TP), which will be considered in the next section. The central difference is that (TP) is stated in temporally sensitive terms, whereas (ITP:O) and (ETP:O) are stated in temporally insensitive terms. Stating (ITP:O) and (ETP:O) without using temporal predications is a requirement of developing the theory of temporal parts on the basis of (4D), which is itself stated in temporally insensitive terms. More generally, temporal insensitivity is a requirement of stating the theory of temporal parts in the spacetime framework.

[8] (T2) is particularly strong, since it allows for radically disconnected extended temporal parts of objects. One might want to impose connectedness as a 'natural' restriction. Since this issue is irrelevant for present purposes, I shall stick to the unrestricted doctrine of arbitrary temporal parts.

that is maximal under some *unity relation*, or *genidentity relation*.[9] There is a unity relation for particles, one for plants, one for persons, etc.; and a person, for instance, is a maximal sum of temporal parts, each of which stands in the unity relation for persons to all the others (and to itself).[10] This construal of ordinary objects is the third thesis:

(T3) An ordinary object is a sum of temporal parts that is maximal under some unity relation.

This completes the theory of temporal parts.

The theory of temporal parts is a development of (4D), the thesis that an ordinary object occupies a unique temporally extended spacetime region. However, (4D) does not entail that ordinary objects have temporal parts. That is, (4D) is compatible with the denial of the doctrine of arbitrary temporal parts, (T2). For one could consistently hold that an ordinary object occupies a single temporally extended spacetime region without having any parts that occupy corresponding temporal parts of that region. Whether or not four-dimensionalism without temporal parts is attractive, it is certainly coherent.

Three-dimensionalism with temporal parts, on the other hand, is not a coherent combination. Assume that *y* is an ordinary object, and that *x* is an instantaneous temporal part of *y*, as defined in (ITP:O). Then *x* is a part of *y*, and *x* occupies a unique instantaneous spacetime region. Presupposing (3D), *y* occupies many instantaneous regions. Since *x* is a part of *y* simpliciter, *x* occupies many instantaneous regions just as *y* does, or else *x* occupies no spacetime region at all. In both cases *x* does not occupy a unique instantaneous region, which is what we assumed. The point is not that *y*'s parts occupy the same regions as *y* does. The point is rather that *y*'s parts occupy multiple regions if *y* does. The only way for a multiply located object *y* to have a part *x* with a unique location is for *x* to be a part of *y* at a single time only. However, temporal parts, as defined in (ITP:O), are parts that objects have simpliciter. A similar clash with (3D) can be shown to arise on the assumption that *x* is an extended temporal part of *y*, as defined in (ETP:O). It follows that (3D) does not allow ordinary objects to have temporal parts.

Temporal counterparts

The *theory of temporal counterparts*, as I call it, is a close variant of the theory of temporal parts. Both theories adopt an ontology of temporally

[9] For the notion of a unity or genidentity relation, see Perry (1975: intro.).

[10] See Lewis (1976: 59).

extended spacetime worms with temporal parts, but differ concerning what counts as an ordinary object, such as a person. According to the theory of temporal parts, a spacetime worm counts as a person, but according to the theory of temporal counterparts, each instantaneous temporal part of a spacetime worm counts as a person, and all the persons that are temporal parts of a spacetime worm are temporal counterparts of each other and of themselves.[11]

The theory of temporal counterparts can be stated more formally on the basis of the theory of temporal parts. Instead of letting theses (T1)–(T3) be about ordinary objects, let them be about what may be called *super-objects*, such as super-persons. Thus, a super-object occupies a unique temporally extended spacetime region—thesis (T1*); a super-object has temporal parts corresponding to temporal parts of the region that it occupies—thesis (T2*); and a super-object is a sum of temporal parts that is maximal under some unity relation—thesis (T3*). Given the notion of a super-object, the notion of an ordinary object may be characterized in terms of it: an ordinary object is an instantaneous temporal part of a super-object. This characterization of an ordinary object must be added as a fourth thesis, (T4*), to (T1*)–(T3*). An immediate consequence of the identification of an ordinary object with an instantaneous temporal part of a super-object is that ordinary objects stand in unity relations to other ordinary objects—this follows from thesis (T3*). These unity relations may be called *temporal-counterpart relations*. Then a super-person is a maximal sum of counterpart-interrelated persons.

The theory of temporal counterparts has the consequence that an ordinary object, being an instantaneous temporal part of an extended spacetime worm, occupies a unique temporally unextended spacetime region. Since this thesis is based on a variant of four-dimensionalism—four-dimensionalism for super-objects—it will be called four-dimensionalism*:

(4D*) (i) an ordinary object occupies a unique spacetime region, and (ii) this spacetime region is temporally unextended.

What (4D*) shares with (3D) is that an ordinary object occupies a temporally unextended spacetime region. The claim that objects lack temporal extension makes (4D*) a variant of three-dimensionalism. What (4D*) shares with (4D) is that an ordinary object occupies a unique spacetime region, that the occupation-relation is one–one. Notice that, while the theory of temporal counterparts entails (4D*), the latter does not

[11] This is what Sider calls the *stage view*. He defends it in Sider (1996) and (2001: 5.7). See also Hawley (2001).

	TTP	TTC	Neither TTP nor TTC
3D	no	no	yes
4D	yes	no	yes
4D*	no	yes	yes

Figure 2

entail the theory of temporal counterparts (as will become clear when we consider presentism below).[12]

To conclude this and the previous subsection, the space of possible combinations may be represented as shown in Figure 2, where 'TTP' refers to the theory of temporal parts and 'TTC' refers to the theory of temporal counterparts.

2.3 ENDURANTISM AND PERDURANTISM

That objects persist is the simple answer to the question how objects are in ordinary time. Three-dimensionalism and four-dimensionalism are answers to the question how objects are in spacetime. How are three-dimensionalism and four-dimensionalism related to the metaphysics of persistence?

I shall begin by examining the standard approach to the metaphysics of persistence. Given that ordinary objects persist, we may ask: *how* do they persist? What is the nature of their persistence? The two rival answers are commonly known as *endurantism* and *perdurantism.*[13] For ease of exposition, let us begin with perdurantism. Objects perdure in virtue of having temporal parts. Temporal parts are parts along the temporal dimension, just as spatial parts are parts along the spatial dimensions. In Section 2.2, the notion of a temporal part of an object was defined at the level of spacetime; having a temporal part was understood in terms of occupying a spacetime region simpliciter and having a part simpliciter. The standard strategy is to define the notion of a temporal part at the level of

[12] By (ITP:O), it trivially follows from (4D*) that an ordinary object has itself as an instantaneous temporal part. This is irrelevant. What matters is that (4D*) entails neither the theory of temporal parts nor the theory of temporal counterparts. The former theory assumes (4D), which is incompatible with (4D*). The latter theory assumes (4D) for super-objects, which clashes with (4D*) if combined with presentism (see below).

[13] The terms 'endurantism' and 'perdurantism' derive from Johnston (1987).

ordinary time; having a temporal part is understood in terms of existence at a time and having a part at a time. My left hand is part of me today, but if I cut it off tomorrow, it is not part of me then. Given parthood at a time, overlap at a time may be defined in terms of it: x overlaps y at $t =_{df}$ something is part of x at t and part of y at t. The notion of a temporal part can now be given the following definition:

(TP) An object x is a *temporal part* of an object y at a time $t =_{df}$ (i) x exists at t, but only at t, (ii) x is a part of y at t, and (iii) x overlaps at t everything that is part of y at t.[14]

With these notions in hand, perdurantism may be stated. Perdurantism is intended as one answer to the question how objects persist. Since persistence is existence at various times, perdurantism may be stated as the following account of how objects exist at a time:

> *Perdurantism*
> Necessarily, an ordinary object x exists at a time t iff x has a temporal part at t.

Next, endurantism:

> *Endurantism*
> Necessarily, an ordinary object x exists at a time t iff x is wholly present at t.

These two theses are rivals only with respect to persisting objects, or continuants—that is, with respect to objects existing at more than one time. For, if an object x exists at a single time t only, then x trivially has itself as a temporal part at t according to (TP). Moreover, x is trivially wholly present at t. Thus, if x is instantaneous, then endurantism and perdurantism come out as compatible. If x exists at various times, however, then, according to perdurantism, x has different proper temporal parts, temporal parts distinct from itself, at each of those times at which x exists. According to endurantism, by contrast, what is present at the various times at which x exists, still assuming that x is a continuant, is not just a temporal part of x, but rather the whole of it.[15]

That ordinary objects persist is a fact of ordinary time. According to standard endurantists, persistence has an underlying nature at the level of ordinary time: objects are wholly present at various times. According

[14] This is Sider's definition, to be found in his (2001: 2.2).

[15] Characterizations of an object as being 'wholly present' can be found in Wiggins (1980), Mellor (1981), Lewis (1986*a*), Simons (1987), Markosian (1994), Zimmerman (1996), Rea (1998), Merricks (1999), Hawley (2001), and Hudson (2001).

to standard perdurantists, persistence has a different underlying nature at the level of ordinary time: objects have different temporal parts at various times. The thesis of temporal supervenience says that all facts about ordinary time logically supervene on facts about spacetime. If we assume that this thesis is true, facts of persistence have an underlying nature at the level of spacetime. It is clear that an object's temporal existence logically supervenes on the object's spatiotemporal location. Three-dimensionalism and four-dimensionalism are rival accounts of an ordinary object's mode of spatiotemporal location. The problem of temporal supervenience thus licenses further accounts of the nature of persistence in terms of three-dimensionalism and four-dimensionalism. An account of persistence at the level of spacetime requires a bridge principle concerning instants, or times. I shall follow standard practice and construe an instant, or time, as a maximal set of simultaneous spacetime points and regions.[16] Spatiotemporal versions of endurantism and perdurantism may then be stated as follows:

> *Spatiotemporal perdurantism*
> Necessarily, an ordinary object x exists at a time t iff x has a temporal part that occupies a member-region of t.
>
> *Spatiotemporal endurantism*
> Necessarily, an ordinary object x exists at a time t iff x occupies a member-region of t.[17]

How are these new versions of endurantism and perdurantism related to the standard versions? Spatiotemporal endurantism and perdurantism are not designed as replacements of standard endurantism and perdurantism. The spatiotemporal versions are independently motivated—owing to the problem of temporal supervenience they are needed anyway—and can

[16] I shall say more about this bridge principle in Sect. 3.1.

[17] I state spatiotemporal perdurantism and endurantism as strict biconditionals in order to render them directly comparable to perdurantism and endurantism in their standard formulations. In the full picture of four-dimensionalist supervenience and three-dimensionalist supervenience to be developed in Sects. 4.1 and 5.1, respectively, these biconditionals will emerge as instances of the principles (TS_{4D}) and (TS_{3D}), respectively, by means of which the following perdurantist and endurantist accounts of the spatiotemporal supervenience base of persistence can be explained:

> *Perdurantist supervenience*
> The fact that an object x exists at a time t is entailed by the fact that x has a temporal part that occupies a member-region of t.
>
> *Endurantist supervenience*
> The fact that an object x exists at a time t is entailed by the fact that x occupies a member-region of t.

in principle coexist with the standard versions. The problem of temporal supervenience, however, is not the only reason for exploring the nature of persistence at the level of spacetime. In fact, there are reasons for preferring to work with the spatiotemporal versions of endurantism and perdurantism instead of the standard versions. I shall focus on endurantism, since the advantages of the spatiotemporal version over the standard version are weightier in the case of endurantism than those in the case of perdurantism.

First, there is reason for doubt that standard endurantism is a sensible thesis. What does it mean for x to be *wholly present* at t? One might, for a start, take 'x is wholly present at t' to mean that everything that is part of x at t exists at t. But then the allegedly controversial thesis that objects are wholly present whenever they exist becomes trivial. For no one, including the friend of temporal parts, would deny that a part of an object at a given time must exist at that time. Alternatively, the sense of 'wholly present' might be defined as follows:

(WP) An object x is *wholly present* at $t =_{\mathrm{df}}$ everything that is at any time part of x exists at t and is part of x at t.[18]

The problem with this definition is that it is too strong. For, if it is plugged into the statement of standard endurantism, the latter entails that it is impossible for an object to gain or lose parts over time. Although mereological essentialism is a view held by some philosophers,[19] merely opposing temporal parts should not commit one to this view. A further approach is to formulate endurantism *via negationis*, as the denial of the possibility of temporal parts, and to drop the problematic concept of being wholly present at a time: necessarily, if an ordinary object x exists at t, then x does not have a temporal part at t. In this clothing, endurantism is clearly a rival of perdurantism, but endurantism is not a positive account of the nature of persistence anymore, which is precisely what endurantism was meant to be. What the standard endurantist promises but struggles to deliver is a condition for existence at a time that is not only necessary but also sufficient.[20]

These considerations show that the task of stating endurantism at the level of ordinary time is difficult. My suspicion is that persistence does

[18] Cf. Sider (2001: 64).

[19] Most notably Roderick Chisholm; see Chisholm (1976: app. B).

[20] For a further problem with the denial of the possibility of (proper) temporal parts, see Sider (2001: 64–5). In the same section, Sider discusses various restrictions on endurantism in the formulation in terms of being wholly present, as defined in (WP), and comes to the conclusion that (standard) endurantism is elusive. For further discussion of the problem of stating endurantism, see Markosian (1994) and Merricks (1999).

not allow for an adequate endurantist account at the level of ordinary time. I suggest that the original endurantist intuition is that instead of an ordinary object's having different temporal parts at the different times of its existence, the object *itself* exists at these different times. Then it turns out that this is just the uncontroversial persistence intuition—an ordinary object exists at different times—an intuition that the perdurantist is well able to accommodate. As a result, the endurantist is pushed to strengthen her thesis, which is when the problematic notion of being wholly present at a time enters the stage. As we just saw, this notion is problematic, since, on the obvious construals of being wholly present, endurantism is either trivial or unacceptably strong. The false assumption in this genesis of the endurantist predicament is that persistence allows for an endurantist account at the level of ordinary time. Accordingly, it is wrong to demand from the endurantist a statement of her position that amounts to such an account. The way forward for the endurantist is to give an account of the nature of persistence at the level of spacetime and to rest content with this account. Spatiotemporal endurantism, as stated above, is a non-trivial picture of how ordinary objects are in spacetime, which mirrors the ordinary fact that an object itself exists at different times, by letting the object itself occupy multiple instantaneous spacetime regions. Moreover, spatiotemporal endurantism avoids the problems with standard endurantism just discussed, because it is a positive, non-mereological account.

The second reason for preferring to work with the spatiotemporal version of endurantism instead of the standard version is more important than the first reason but can only be foreshadowed at this stage of the enquiry. Change presupposes persistence; for an object to be different at different times, the object must exist at these times. Accordingly, the important task of exploring the nature of change depends to some extent on insights regarding the nature of persistence. Now, even if an acceptable statement of endurantism at the level of ordinary time were available, the statement of endurantism in terms of three-dimensionalism promises an explanation of the nature of change that is available only at the level of spacetime, and hence would not be available on the mere assumption of standard endurantism. The explanation in question is the *temporal-regions account* of temporal supervenience, to be put forth and defended in Chapter 5. By the end of that chapter, I will have reached the conclusion that the temporal-regions account offers the best metaphysic of persistence and change. It is primarily for this reason that I will henceforth focus on spatiotemporal endurantism and perdurantism and mean these versions when I speak of endurantism and perdurantism simpliciter.

2.4 ETERNALISM AND PRESENTISM

How are three-dimensionalism and four-dimensionalism related to eternalism and presentism? While the former distinction concerns the way in which objects are in time, the latter distinction concerns the ontology of time itself. Eternalism and presentism are different views on the status of the present. Eternalism is the view that non-present times, past and future ones, are as real as the present time, and no time is special. Presentism is the view that the present is special: only the present time is real, and everything that exists, exists now. Saying that only the present is a real time leaves room to say that there exist ersatz times that are distinct from the present and somehow constructed out of presently existing material (more on ersatz times in Section 4.2). As pointed out in Section 1.5, the notions of past, present, and future presuppose the distinction between time and space. Since this distinction disappears in the spacetime conception, the notions of past, present, and future are confined to the ordinary conception of time; these notions are not defined for spacetime. Eternalism and presentism are therefore features of ordinary time, not of spacetime.

Nonetheless, it is possible to formulate features of spacetime that mirror to some extent the standard eternalism–presentism distinction. In order to avoid introducing new terms, I shall distinguish between eternalism and presentism about ordinary time—the views just stated—and, somewhat inaccurately, eternalism and presentism about spacetime. Eternalism and presentism about spacetime may be stated as follows:

(E) There is a manifold of hyperplanes of simultaneity.
(P) There is only one hyperplane of simultaneity.

Note that (P) is not the only possible spacetime formulation of presentism. But it seems to be the natural view for a presentist about spacetime to adopt.[21] Note further that (P) looks like a coherent view only as long as we presuppose classical spacetime. As soon as we take into account relativistic spacetime, however, (P) is threatened with inconsistency. The reason is that (P) in effect selects an arbitrary frame of reference to single out an absolute relation of simultaneity. Correspondingly, (P) implies that there is a fact of the matter as to which events on Saturn are simultaneous with any event here on Earth. Special Relativity, however, denies that there are any such facts. In Minkowski spacetime (see Section 1.6), there is

[21] For a discussion of various spacetime formulations of presentism, see Sider (2001: 2.4).

no observer-independent notion of simultaneity. Hence, presentism about spacetime is scientifically revisionist.[22]

Let us now examine the relationship between the spatiotemporal theses of three-dimensionalism, four-dimensionalism, eternalism and presentism—(3D), (4D), (E), and (P). Once these relationships are clear, theses about ordinary time may be brought into the picture. First, (3D) is incompatible with (P) and thus presupposes (E). It is easy to see why. According to (3D), ordinary objects occupy multiple instantaneous spacetime regions, each of which lies on a different hyperplane of simultaneity. This contradicts (P), according to which there is only one such hyperplane. Secondly, (4D) is just like (3D): (4D) is incompatible with (P) and thus presupposes (E). According to (4D), ordinary objects occupy temporally extended regions. Such a region is constituted by a multitude of instantaneous regions, each of which lies on a different hyperplane of simultaneity. According to (P), however, there is only a single hyperplane.

So neither (3D) nor (4D) is available to the presentist about spacetime. How are things in time, then, if presentism about spacetime is true? The presentist says that there is only one hyperplane. Still assuming that ordinary objects are spatiotemporally located, the presentist must therefore say that an ordinary object occupies an instantaneous spacetime region—the kind of region that lies on a hyperplane—and that an object occupies a unique such region, since the presentist does not want to say that an object occupies different simultaneous regions. The resulting view was earlier stated as four-dimensionalism* (4D*). It remains to be pointed out that (4D*) does not entail (P), since (4D*) may be combined with (E) in the form of the theory of temporal counterparts. The space of possible combinations of theses about spacetime may then be represented as shown in Figure 3.

	E	P
3D	yes	no
4D	yes	no
4D*	yes	yes

Figure 3

[22] Craig (2001) claims that facts about absolute simultaneity have a place in SR after all, if the spacetime interpretation of SR is rejected in favour of a suitable 'neo-Lorentzian' interpretation. For discussion, see Balashov and Janssen (2003).

How do eternalism and presentism about ordinary time fit into this picture? First, eternalism about ordinary time. The eternalist says that past and future times are as real as the present time; no time is special. This view is naturally combined with eternalism about spacetime (E), by means of construing the eternalist's *real times* as maximal sets of simultaneous spacetime points and regions—that is, as hyperplanes of simultaneity. Then a manifold of times brings with it a manifold of hyperplanes.

Next, presentism about ordinary time. The standard presentist about ordinary time is a tenser, holding that grammatical tense is semantically primitive and that ordinary time is A-time (see Sections 1.1 and 1.2). It was shown in Section 1.5 that the tenser is unable to accommodate truths about spacetime, whether eternalism about spacetime or presentism about spacetime is presupposed; A-time is incompatible with spacetime. It follows that tenserist presentism is incompatible with both eternalism and presentism about spacetime. There is, however, an alternative version of presentism about ordinary time. The presentist about ordinary time may treat tense in detenserist fashion as predicates of times, but construe times not as real times, but rather as *ersatz times*. These ersatz times may be understood as different abstract representations of a spacetime that contains no more than a single hyperplane; and one ersatz time is special in virtue of representing the instantaneous spacetime correctly. In this case, presentism about ordinary time is compatible with presentism about spacetime (P).[23]

Finally, how do endurantism and perdurantism fit into the picture? According to the (spatiotemporal) endurantist, three-dimensionalist facts about spatiotemporal occupation underlie ordinary facts of persistence. According to the (spatiotemporal) perdurantist, four-dimensionalist facts about spatiotemporal occupation and temporal parts underlie ordinary facts of persistence. As we just saw, three-dimensionalism and four-dimensionalism are incompatible with presentism about spacetime, and therefore presuppose eternalism about spacetime. As a consequence, endurantism and perdurantism presuppose eternalism about spacetime. Moreover, the eternalist about spacetime is committed to eternalism about ordinary time; a temporally extended spacetime leaves no room for a privileged present. Therefore, the endurantist and the perdurantist are committed to eternalism about ordinary time. This last result goes against the view held by some philosophers that endurantism is incompatible with eternalism about ordinary time and thus presupposes presentism about ordinary time.[24] The result further goes against the claim that perdurantism

[23] I will look at the combination of presentism about ordinary time and presentism about spacetime in Sect. 4.2.

[24] See Carter and Hestevold (1994), Markosian (1994), and Merricks (1995, 1999).

is compatible with presentism about ordinary time.[25] To repeat, the incompatibility of endurantism and perdurantism with presentism about ordinary time hangs on the construal of endurantism and perdurantism as theses about the spatiotemporal facts that ground persistence. This result remains unaltered by the fact that the philosophers who hold the above-mentioned claims about endurantism, perdurantism, and presentism work with the standard versions of endurantism and perdurantism. The reason why the result remains unaffected is that the standard versions of endurantism and perdurantism, if they are acceptable, form a continuum with the spatiotemporal versions. There is no escape from temporal supervenience.

[25] See Sider (2001: 2.4).

3

Temporal Predication and Supervenience Failure

The problem of temporal supervenience is to explain how facts of persistence and temporal instantiation, which are facts about ordinary time, supervene on facts about spacetime. Part of the problem of temporal supervenience is the problem of temporal predication, the task of giving an account of the logic and semantics of the language in which the facts of ordinary time are stated. Having rejected tenserism in favour of detenserism in Chapter 1, the problem of temporal predication reduces to the task of giving an account of the logical and semantic function of the modifier 'at *t*' in '*a* is F at *t*', since according to the detenser modification by 'at *t*' underlies ordinary forms of temporal modification such as tense and temporal adverbials. In the project of explaining temporal supervenience, an account of temporal predication functions as an 'analysis' of ordinary temporal facts, an analysis that is meant to build an explanatory bridge from these ordinary temporal facts to their spatiotemporal supervenience base. In this chapter, I shall discuss various accounts of temporal predication that have in common the fact that temporal supervenience cannot be explained on the basis of them. In other words, the accounts to be discussed allow no plausible explanatory link between facts about ordinary time and any facts about spacetime.

3.1 THE RELATIONAL ACCOUNT

On the surface, a temporal predication '*a* is F at *t*' is formed from an atemporal predication '*a* is F' by adding the modifier 'at *t*'. Let us stay with these atemporal predications for a moment and formulate an account of the logical form and semantics for them first. In general, to specify the logical form of an ordinary English sentence is to associate the English sentence with a sentence of a formal language, with the purpose of elucidating the structure of the English sentence and to allow for a clear semantic treatment. The semantics of an ordinary sentence will then be the semantics of its associated formal sentence. The logical form of the atemporal monadic predication '*a* is F', where the copula is tenseless, is straightforward: the sentence contains a one-place predicate 'F( )' and a singular subject term '*a*', yielding: 'F(*a*)'. Similarly for dyadic predications: '*a* is R to *b*' becomes 'R(*a*, *b*)'. Assuming that the semantics of natural language takes the form of a T-theory, the semantics of temporally unmodified monadic and dyadic predications is given by the following theorem:

(T_0) 'F(*a*)' is true $\equiv$ F(*a*)
'R(*a*, *b*)' is true $\equiv$ R(*a*, *b*)

By specifying the literal truth conditions of the sentence on the left-hand side, the right-hand side of the theorem delivers the semantic content of the left-hand side.

Let us move on to temporal predication. The problem of giving an account of the logical and semantic function of 'at *t*' in '*a* is F at *t*'—the problem of temporal predication—may be viewed as the question of how to extend the simple semantic picture of atemporal predications to temporal predications. There are two strategies of approaching this task. The first strategy is to explain away the modifier 'at *t*', and thereby to reduce temporal predications to atemporal predications, a semantic account of which is given by (T_0). The second strategy is to view the modifier 'at *t*' as untouchable, and to modify (T_0) to yield a semantic account of irreducibly temporally modified predications. Both strategies will be considered in detail as we go along.

The simplest way of implementing the first strategy is the *relational account* of temporal predication. The core of this account is the following thesis about logical form: at the surface level the sentence '*a* is F at *t*' contains a temporal modifier 'at *t*' and a one-place predicate 'is F', whereas at the level of logical form the sentence contains no temporal modifier and no one-place predicate, but rather a two-place predicate 'F(,)', where the first place is filled by a term designating an object and the second place is filled by a term designating a time, yielding:

(1) F(*a*, *t*)

As a consequence, apparently monadic temporal predications are really dyadic atemporal predications, a semantic account of which is provided by the dyadic clause of (T_0). This simple logical move pays the price of diverging from surface form. Where we thought that '*a* is round at *t*' and '*a* is red at *t*' contained one-place predicates, they really contain two-place predicates. To put the point in property-speak, where we thought that shapes and colours were properties of objects at times, they turn out to be relations between objects and times. Hence the label 'relational account'.[1]

The relational account is usually discussed in the context of the problem of change. In Section 1.4, I argued that this alleged problem has a semantic side and a metaphysical side. The semantic problem of change is essentially the constraint that an account of temporal predication rule out that a report of change—for instance, 'Zoe is happy at t_1 and Zoe is unhappy at t_2'—imply a contradiction, in this case 'Zoe is happy and Zoe is unhappy'. The relational account satisfies this constraint, because the mentioned report of change is regimented as 'Happy(Zoe, t_1) & Unhappy(Zoe, t_2)', which is obviously free of contradiction.

The idea that temporal predications contain predicates with a time-place, and its consequence that apparent properties such as shapes and colours are really relations between objects and times, have been much criticized. I find most published objections unconvincing.[2] Since my aim is not to defend the relational account, there is no need for an extensive review of this criticism. Here is an example of an influential but weak objection. David Lewis claims that shapes are obviously intrinsic to the objects that have them and therefore cannot be relations between objects and times. What is meant by 'intrinsic'? An intrinsic property is, roughly, a property that could belong to something that was the only object in the universe, lonely or unaccompanied by any contingent object wholly distinct from itself.[3] Since we may picture, say, a ball as spherical while alone in the universe,

[1] It is possible to derive from 'F(a, t)' a sentence containing a one-place predicate by means of the principle of *predicate abstraction*. This principle says that we can form a complex predicate from any open sentence. For instance, 'Zoe is happy' is equivalent to 'Zoe is such that she is happy', which contains a complex predicate 'is such that she is happy'. In general, for any sentence 'S(a)',

(PA) $(S(a) \equiv \lambda x[S(x)](a))$

Applying (PA) to 'F(a, t)' yields the equivalent '$\lambda x[F(x, t)](a)$', which contains the complex one-place predicate '$\lambda x[F(x, t)](\ )$'. In property-speak, the point is that we can turn a statement ascribing a dyadic relation to objects and times into a statement ascribing a relational property to objects, where a relational property is a property that is 'based on' some relation in the sense that the predicate of the relational property is obtained from the predicate of the relation by abstraction. This equivalence is worth mentioning because the relational account is sometimes presented in terms of relational properties. Predicate abstraction is also helpful in clarifying the elusive talk of 'time-indexed properties' occasionally encountered in the literature. The equivalence, however, plays no role in the present discussion. (PA), on the other hand, will soon prove to be important (see Sect. 5.5).

[2] See Lewis (1986*a*: 203; 1988: 87), Johnston (1987: 113), Hawley (1998), Mellor (1998: 8.6), Sider (2001: 4.6, 4.7), and Rodriguez-Pereyra (2003).

[3] This is essentially Jaegwon Kim's definition (1982). For various qualifications, see Langton and Lewis (1998). These qualifications play no role here.

shapes are intrinsic properties by this definition. But the intuition that a ball may be spherical unaccompanied by other physical objects merely shows that shapes are not relations between physical objects. The intuition fails to show that shapes are not relations between physical objects and times. The intuitive test carries some weight if it asks us to cut away all the concrete things that are disjoint from the ball from the universe. The test carries no weight if it asks us to cut away all the times from the universe (along with all the concrete things disjoint from the ball). We do not have a direct intuition about whether or not a given object could have a shape in a timeless universe. Without a pre-theoretic grasp of what times are, a pre-theoretic grasp of the state of objects unaccompanied by times is out of the question. The relationalist has therefore no reason to worry about issues of intrinsicness.

A further charge to the effect that the relational account is counter-intuitive concerns change. It has been argued that the relationalist cannot take seriously the intuition that change is temporal property-variation. It is perfectly commonplace to understand an object *a*'s changing in shape as the variation of *a*'s shapes over time—that is, as *a*'s having different shapes at different times. Since temporal property-variation requires instantiation of a property at a time, the variation-intuition disparages the relational account. For, according to the latter, nothing has a shape at a time, and hence the shapes of things do not vary over time; things have their shape-relations simpliciter. To put the point differently, change is a matter of gaining and losing properties relative to different times. Change is not a matter of simply standing in various relations to different times. Time is the arena of change, not a subject of it. Of course, the relationalist may define change in shape as bearing different shape-relations to different times. But this would not be change as commonly understood. On the ordinary conception of change, nothing changes, strictly speaking, if the relational account of temporal predication is correct.[4]

In response to the charge that the relational account fails to capture the variation-intuition, the relationalist may say that the variation-intuition is a product of the surface form of temporal predications. This way the relationalist is able to explain where the variation-intuition comes from. In a sense, then, the relationalist can capture the variation-intuition. What the relationalist cannot do is take the variation-intuition seriously. Since the characteristic feature of the relational account is that it assigns temporal predications a logical form that diverges significantly from their surface form, the relationalist is bound to play down the variation-intuition along

[4] Objections along those lines are made in Lowe (1988: 73–4), Haslanger (1989*a*: 119–20), and Hinchliff (1996: 120–1).

with our ordinary conception of change as insignificant and misleading. According to the relationalist, it is not strictly and literally true that objects vary in their properties over time. As in the case of all ordinary intuitions, it is controversial whether the variation-intuition should be taken seriously. As an optimist, I find the methodological stance most plausible that attempts to capture our intuitions where possible. From this perspective, an account of logical form that stays close to the surface form of temporal predications is to be preferred to the relational account. My reasons for rejecting the relational account, however, lie deeper.

3.2 RELATIONAL SUPERVENIENCE FAILURE

The demand for an account of temporal predication is part of the broader quest for an account of temporal supervenience. An account of temporal supervenience is a specification of the spacetime facts on which certain types of ordinary temporal fact supervene and an explanation of how they supervene. The ordinary temporal facts at the centre of our attention are facts of persistence and of temporal instantiation. Since reports of persistence and of temporal instantiation are temporal predications, an account of temporal predication plays a significant role in explaining both persistence supervenience and temporal-instantiation supervenience. In this section, I shall argue that the relational account of temporal predication makes a plausible account of persistence supervenience and of temporal-instantiation supervenience unlikely.

Persistence supervenience

Persistence logically supervenes on spatiotemporal location; no two logically possible worlds differ in their facts of spatiotemporal location without differing in their facts of persistence; the facts of spatiotemporal location entail the facts of persistence. Different accounts of the spatiotemporal location of ordinary objects thus constitute alternative options for the supervenience base of the persistence of ordinary objects. The main options, as developed in Chapter 2, are three-dimensionalism and four-dimensionalism. To remind us:

(3D) (i) an ordinary object occupies multiple spacetime regions, and (ii) these spacetime regions are temporally unextended, or instantaneous, and non-simultaneous.

(4D) (i) an ordinary object occupies a unique spacetime region, and (ii) this spacetime region is temporally extended.

Given these accounts of spatiotemporal location, we may distinguish between an endurantist account of persistence supervenience that attempts to explain how the facts of persistence supervene on three-dimensionalist facts about spatiotemporal location, and a perdurantist account of persistence supervenience that attempts to explain how the facts of persistence supervene on four-dimensionalist facts about spatiotemporal location. Our question is whether any of these accounts of persistence supervenience is possible if the relational account of temporal predication is correct.

Let us begin with three-dimensionalism (3D). If the persistence of ordinary objects supervenes on three-dimensionalist facts about spatiotemporal location, then this supervenience is not a brute fact and thus requires explanation. How can it be explained that the fact that an ordinary object occupies various instantaneous spacetime regions entails the fact that this object persists, that it exists at various times? The relational account of temporal predication says that a report of persistence '*a* exists at t_1 and *a* exists at t_2' is to be regimented as 'E(a, t_1) & E(a, t_2)', where 'E(,)' is a two-place existence-predicate. It is clear that this thesis about logical form on its own does nothing to clarify the link between *a*'s spatiotemporal location and *a*'s persistence. The relational account thus needs to be combined with some bridge principles.

The first bridge that is required concerns the relationship between instants, such as t_1 and t_2, and spacetime points and regions. This bridge is delivered by the natural construal of instants as maximal sets of simultaneous spacetime regions. Linking instants with spacetime regions in this way raises a question, though. In Section 1.3, reductionism was rejected on the ground that the reductionist makes the implausible claim that we have all along been speaking the language of physics without noticing. Why is the identification of the instants of ordinary time with sets of spacetime regions not an objectionable instance of reductionism? Compare the case of instants to the following: heat is mean kinetic energy, whereas the concept of heat is distinct from the concept of mean kinetic energy. So when we are speaking about heat we are, in a sense, speaking about energy. But, since we are using a different concept, we are not speaking the language of physics. Analogously for instants: instants are hyperplanes, whereas the concept of an instant is distinct from the concept of a hyperplane. So, when we are speaking about instants, we are, in a sense, speaking about hyperplanes. But, since we are using a different concept, we are not speaking the language of physics. This is why the identification of instants with hyperplanes does not amount to reductionism.

Is the construal of instants as hyperplanes sufficient to bridge the gap between *a*'s spatiotemporal location and *a*'s persistence? The answer is 'no'. For it seems perfectly possible for *a* to bear the occupation-relation to various instantaneous spacetime regions without bearing the existence-relation to any sets of instantaneous spacetime regions. Life would be easy for the relationalist if the existence-relation could simply be identified with the occupation-relation. But this identification is out of the question, since the occupation-relation is a relation between objects and spacetime regions, whereas the existence-relation is a relation between objects and times—that is, sets of spacetime regions. So there are different kinds of facts in play, occupation facts and existence facts, which prima facie seem to have little connection. A big conceptual gap remains.

A straightforward strategy to bridge the gap lies in defining the meaning of the existence-predicate in terms of the occupation-predicate. This may be done in the following way:

(L1) $\mathrm{E}(a, t) =_{\mathrm{df}} \exists R(R \in t \mathbin{\&} \mathrm{O}(a, R))$

Intuitively, that *a* bears the existence-relation to *t* means that *a* occupies some member-region of *t*. (Note that, since the occupation-predicate is more basic than the existence-predicate and was assumed to be a primitive predicate (see Section 2.1), the definition cannot run the other way around: $\mathrm{O}(a, R) =_{\mathrm{df}} \exists t(R \in t \mathbin{\&} \mathrm{E}(a, t))$.) (L1) delivers the link that the relationalist requires. Assuming that *a* occupies various instantaneous spacetime regions, and given that each of these regions is a member of a maximal set of instantaneous regions, the latter being times, (L1) entails that *a* bears the existence-relation to various times. Hence, three-dimensionalism entails persistence in the relationalist's sense.

There is a variant of this endurantist explanation of persistence supervenience that starts by identifying an instant not with a hyperplane—that is, with a maximal set of instantaneous points and regions—but rather with the sum of all points and regions on a hyperplane. This has the consequence that the relationalist's existence-relation is now a relation between objects and big spacetime regions, just as the occupation-relation is a relation between objects and regions. Given this construal of instants, the meaning of the existence-predicate may be defined as follows:

(L2) $\mathrm{E}(a, t) =_{\mathrm{df}} \exists R(\mathrm{P}(R, t) \mathbin{\&} \mathrm{O}(a, R))$

where '$\mathrm{P}(x, y)$' is to be read as '*x* is a part of *y*'. Intuitively, that *a* bears the existence-relation to *t* means that *a* occupies some subregion of *t*. Just as (L1), (L2) is sufficient to explain the supervenience of temporal existence on spatiotemporal occupation. Assuming that *a* occupies various instantaneous spacetime regions, and given that each of these regions is a part of a maximal sum of instantaneous regions, the latter being times, (L2) entails that *a* bears the existence-relation to various times.

Definitions (L1) and (L2) share the same defect; they are too strong. In Section 1.3, I characterized reductionism as a view about the relationship between the conceptions of ordinary time and spacetime, according to which these conceptions are really just one conception, because the ordinary conception is reducible to the spacetime conception. To reduce the ordinary conception to the spacetime conception is to define the meanings of ordinary temporal terms in spacetime terms. Thus, to define the meaning of, or concept associated with, a predicate belonging to the ordinary temporal scheme in terms of a predicate belonging to the spacetime scheme in the way exemplified in (L1) and (L2) is a reductionist move: to say that an object exists at a certain time really means that the object occupies a spacetime region that lies on a certain hyperplane of simultaneity, and hence the existence-predicate expresses a spatiotemporal concept. What is objectionable about such meaning postulates is their violent assimilation of what appear to be fundamentally different conceptions—the claim that our naive conception of ordinary time, which distinguishes times from places, is the highly developed conception of spacetime in disguise. An optimistic stance that recognizes ordinary time and spacetime as distinct conceptions is to be preferred.

It would not be reductionist for the relationalist to say that the existence-predicate expresses an ordinary concept that picks out a relation that an object a bears to a time t just in case a occupies a member-region or a subregion of t. That is, the relationalist needs one of the following principles that are weaker than (L1) and (L2):

(L1*) $\Box[\mathrm{E}(a, t) \equiv \exists R(R \in t \,\&\, \mathrm{O}(a, R))]$
(L2*) $\Box[\mathrm{E}(a, t) \equiv \exists R(\mathrm{P}(R, t) \,\&\, \mathrm{O}(a, R))]$

The question that opens the conceptual gap that threatens the relational account is why it is not possible for an object to occupy a member-region or a subregion of t without existing-at t. A substantive answer to this question is to claim that the existence-predicate means what (L1) or (L2) says it means. But it is not a substantive answer to say that, necessarily, an object exists-at t iff it occupies a member-region or a subregion of t. For the same question arises: why is this necessary? Analogously, in order to explain why water-facts supervene on H_2O-facts, it is not enough to say that, necessarily, something is water iff it is H_2O. An explanation of this kind of supervenience is naturally based on a functional analysis of the concept 'water'. Given such an analysis, we can say that water-facts supervene on H_2O-facts because H_2O plays the role captured by the ordinary concept 'water'. No such functional analysis seems to be possible in the case of the concept 'existence'. Can 'existence' be analysed in another, non-functional and non-reductionist, way that links existence facts with occupation facts?

I cannot see how. Therefore I cannot see how the relationalist might close the conceptual gap. My view is that the only sensible way of explaining the supervenience of facts of temporal existence on facts of spatiotemporal location is via a semantic analysis of the modifier 'at t' (such an analysis will be discussed in Sections 4.1 and 5.1). Since the relational account logically analyses such modifiers away, this route is blocked. A plausible endurantist account of persistence supervenience therefore seems unlikely if the relational account of temporal predication is correct.

For much the same reasons, the relationalist is incapable of giving a perdurantist account of persistence supervenience, or indeed any account at all. No matter what kind of account of persistence supervenience is attempted, the relationalist still has the problem that the existence-relation cannot be identified with the occupation-relation, and she still seems to have only one way of bridging the gap between occupation facts and existence facts, namely via defining the meaning of the existence-predicate in terms of the occupation-predicate. How exactly this is done changes with respect to which account of spatiotemporal occupation is chosen as supervenience base. What remains in each case is the charge that the meaning of, or concept associated with, the existence-predicate cannot be defined in terms of the occupation-predicate, because the existence-predicate and the occupation-predicate belong to different conceptual schemes. If four-dimensionalism is chosen as supervenience base, the following definition may be given:

(L3) $\mathrm{E}(a, t) =_{\mathrm{df}} \exists R \exists R' (R \in t \ \&\ \mathrm{P}(R, R') \ \&\ \mathrm{O}(a, R'))$

Intuitively, that a bears the E-relation to t means that a occupies a region that has a part that is a member of t (assuming that t is a hyperplane).[5] (L3) has the consequence that four-dimensionalism entails persistence, since four-dimensionalism says that an object occupies an extended region that has instantaneous regions as parts that are members of times. (L3), however, is an objectionable instance of reductionism, as is any definition of the existence-predicate in terms of the occupation-predicate. The result is that, if the relational account of temporal predication is correct, then persistence supervenience remains unexplained.

[5] (L3) is a definition of the existence-predicate in terms of parts of the region occupied by an object. Alternatively, the existence-predicate may be defined by a four-dimensionalist in terms of regions occupied by temporal parts of objects:

(L3′) $\mathrm{E}(a, t) =_{\mathrm{df}} \exists R \exists x (R \in t \ \&\ \mathrm{TP}(x, a) \ \&\ \mathrm{O}(x, R))$

where 'TP(x, y)' means that x is a temporal part of y. Intuitively, that a bears the existence-relation to t means that a has a temporal part that occupies a member-region of t. (L3′) is objectionable on the same grounds as (L3).

Temporal-instantiation supervenience

Ordinary facts of temporal instantiation, facts of having a property at a time, logically supervene on a certain type of fact about spacetime. The problem of temporal-instantiation supervenience is to determine which type of spacetime fact ordinary temporal-instantiation facts supervene on, and to explain how they supervene. According to the relational account of temporal predication, the ordinary facts of temporal instantiation are disguised relational facts: apparent monadic properties instantiated at a time are really dyadic relations to a time instantiated simpliciter; analogously for any apparent *n*-adic relations instantiated at a time. On which spacetime facts do these ordinary relational facts supervene? Focus on the example of shape-relations between objects and times. It is clear that the relationalist cannot account for the instantiation of shape across spacetime in terms of monadic shapes. This would immediately cut any explanatory link to ordinary relational facts that involve shape-relations, since shape-properties and shape-relations do not seem to have much in common. So the shapes instantiated in spacetime must be shape-relations. Relating what? Since the concept of a time is not part of the spacetime scheme, the relationalist cannot just carry over her account of ordinary temporal instantiation, according to which shapes relate objects and times. The relationalist needs a bridge principle identifying the times of ordinary time with certain entities in spacetime. She then needs to construe shape-relations in spacetime as relations between objects and these entities. Since, as we already know, the natural construal of times is as maximal sets of simultaneous spacetime regions, the relationalist is committed to the view that shapes are distributed across spacetime in virtue of being relations between objects and maximal sets of simultaneous spacetime regions. One might find it more natural to construe shape-relations as relations between objects and spacetime regions, as opposed to sets of regions. This construal, however, fails to deliver the required link from spacetime facts to ordinary facts, since bearing a shape-relation to a region does not entail bearing a shape-relation to a time. By the bridge principle identifying times with hyperplanes, an object's bearing a shape-relation to a hyperplane entails the object's bearing a shape-relation to a time. Supervenience has thus been explained.

The relationalist's account of the supervenience base of ordinary facts of temporal instantiation also solves the metaphysical problem of change (as characterized in Section 1.4). Since times have no place in spacetime, the instantiation of properties in spacetime cannot be relativized to times. Then the problem arises how to account for the supervenience base of ordinary facts of change. An object *a*'s having incompatible shapes at different times surely cannot be grounded in *a*'s having incompatible shapes simpliciter. The metaphysical problem of change thus poses the task of avoiding a threat of contradiction. The relationalist avoids contradiction by saying that shapes are distributed across spacetime in virtue of being relations between objects and hyperplanes. So objects may occupy distinct instantaneous regions *R* and *R*′, which accords with the three-dimensionalist position on spatiotemporal location, and bear incompatible shape-relations to the hyperplanes that have *R* and *R*′ as members. Given that times are hyperplanes, an object *a*'s bearing incompatible shape-relations to different times is entailed by *a*'s bearing incompatible shape-relations to different hyperplanes.

The first critical point to make about the relational account of temporal-instantiation supervenience is that it is oddly complex to bring in maximal sets (or maximal sums) of regions to explain spatiotemporal instantiation. Owing to this complexity, one gets the feeling that the relational account of spatiotemporal instantiation is a simple projection from the relational account of temporal instantiation at the level of ordinary time, and hence that spatiotemporal instantiation is explained in terms of ordinary temporal instantiation—that the shape of spacetime is explained in terms of the shape of ordinary time—and not the other way around, as it should be done.

A further worry about the relational account of temporal-instantiation supervenience arises from a plausible requirement on temporal predication, what I will call the *temporal-existence requirement.* It seems obvious that something can have a property at a time only if it exists at a time.[6] This is the temporal variant of the atemporal principle that existence is required for instantiation. Assuming temporal supervenience, not only ordinary facts of temporal instantiation and temporal existence supervene on spacetime facts, but the logical relations of ordinary instantiation facts and existence facts supervene on the logical relations of certain spacetime facts as well. Given that *a*'s being F at a time implies that *a* exists at a time, whatever spacetime fact entails that *a* is F at a time, this spacetime fact should imply whatever spacetime fact entails that *a* exists at a time. Given the plausible assumption that how an object is in ordinary time supervenes on how the object is in spacetime, and hence that ordinary facts of temporal existence supervene on facts of spatiotemporal location, it follows that, whatever spacetime fact entails that *a* is F at a time, this fact of spatiotemporal

[6] It is a stronger requirement that something can have a property at a time only if it exists at *that* time. I tend to hold this stronger version as well, although there are well-known prima facie counter-examples, such as that Socrates can be admired now without existing now.

instantiation should imply that *a* is spatiotemporally located.[7] This is how the temporal-existence requirement translates to the level of spacetime, forming a constraint on an account of spatiotemporal instantiation.

The problem for the relational account of spatiotemporal instantiation is that it violates the temporal-existence requirement or at least covers it in darkness. For the relationalist's supervenience base of *a*'s being F at a time fails to imply that *a* is spatiotemporally located. Since *a*'s being spatiotemporally located is what underlies *a*'s existing at a time, the relationalist fails to capture the temporal-existence requirement that *a* is F at a time only if *a* exists at a time. The problem is that it seems possible for an object to bear a shape-relation to a set of spacetime regions without occupying any portion of spacetime. This seems possible because there is nothing in the relationalist's shape-relations that could explain a necessary connection with the occupation-relation. If there were a necessary connection of this sort, it would be a complete mystery at the heart of the relational account. My view is that the only sensible way of capturing the temporal-existence requirement at the level of spacetime is via a semantic analysis of the modifier 'at *t*' (see Sections 4.1 and 5.1). Since the relational account logically analyses such modifiers away, no sensible explanation of the temporal-existence requirement seems to be available.

I conclude that a workable relationalist account of persistence supervenience and of temporal-instantiation supervenience is unlikely. This failure to allow for an explanation of temporal supervenience constitutes a significant flaw of the relational account of temporal predication.

3.3 THE INTENSIONAL ACCOUNT

The relational account represents the simplest way of explaining away temporal modification from '*a* is F at *t*', and hence of reducing temporal predications to atemporal predications, a semantic account of which is provided by (T_0). Reducing temporal modifiers is one strategy of tackling the problem of temporal predication, and we will look at another implementation of this strategy in Chapters 4 and 5. But now let us consider the second strategy: to leave temporal modifiers alone and to modify (T_0) instead.

[7] The ability of the relationalist to explain the supervenience of temporal existence on spatiotemporal location was questioned in the previous subsection. In the present context, I am ignoring this complication for the relationalist, in order to state a different problem.

Our question is how to extend the simple semantic picture of temporally unmodified predications given in (T_0) to temporally modified predications. On the surface, the sentences '*a* is F at *t*' and '*a* is G to *b* at *t*' contain the modifier 'at *t*'. The relational account of temporal predication is the thesis that the logical form of these sentences diverges from their surface form in virtue of containing no such modifier. As an alternative, we may consider a view according to which the logical form of temporal predications mirrors their surface form. This is the view that 'at *t*' in '*a* is F at *t*' functions logically as a sentence modifier, 'At *t*', attaching to an atemporal predication 'F(*a*)' to form a temporal predication

(2) At t[F(a)][8]

Having recognized a temporal sentence modifier 'At *t*' at the level of logical form, the question of the semantic interpretation of 'At *t*' arises. Here we can distinguish between the view that 'At *t*' is semantically reducible and the view that 'At *t*' is semantically primitive. Recall from the discussion of tense in Section 1.1 that a temporal modifier is semantically irreducible if the modifier cannot be understood in other terms and is used in the metalanguage in which the truth conditions of sentences in which the modifier occurs are stated. A modifier is semantically reducible if it does not recur in the truth conditions of sentences in which the modifier occurs. The relational account views apparent temporal modifiers as semantically reducible. According to the relationalist, the modifiers disappear at the level of logical form, which trivially implies that they do not occur in the truth conditions. The view to be considered here is that temporal modifiers are semantically irreducible. Given that 'at *t*' functions as a sentence modifier, the truth conditions of monadic and dyadic temporal predications may be stated as follows:

(T_1) 'At t[F(a)]' is true $\equiv$ At t[F(a)]
'At t[R(a, b)]' is true $\equiv$ At t[R(a, b)]

[8] Once we have a sentence modifier 'At *t*', we can derive a predicate modifier from it by means of the predicate abstraction principle (PA). Letting 'S(*a*)' be 'At *t*[F(*a*)]', we get from (PA):

$$\Box[\text{At } t[F(a)] \equiv \lambda x[\text{At } t[F(x)]](a)]$$

Here the modifier 'At *t*' has two different functions: on the left-hand side it functions as a sentence modifier, whereas on the right-hand side it functions as a predicate modifier. Since the predicate-modifier reading of temporal predications is derived from the sentence-modifier reading, the former does not require separate attention. (I say *the* predicate-modifier reading and *the* sentence-modifier reading, because I take it that there are not various ways for 'at *t*' to be a predicate modifier or a sentence modifier. There is a single logical function of predicate modifier and a single logical function of sentence modifier, though there may be various formal representations of each of these functions.)

In a theory of meaning the truth conditions of sentences function as theorems that give the semantic properties of these sentences. In addition to these theorems, a theory of meaning consists of axioms that give the semantic properties of words and their modes of combination, from which the theorems that give the semantic properties of sentences can be deduced. In order to understand how the semantic values of sentences depend on the semantic values of their parts, we therefore need to have a look at some semantic clauses for subsentential expressions. The reason for doing so in the present context is that the semantic mechanism underlying (T_1) is both the central and most problematic feature of the account of temporal predication under consideration.

The theorems for temporally unmodified monadic and dyadic predications have the following form:

(T_0) 'F(a)' is true $\equiv$ F(a)
'R(a, b)' is true $\equiv$ R(a, b)

Here is how these theorems may be deduced from simple axioms. The semantic value of a name is its referent. The referent of 'a' is a and the referent of 'b' is b:

(3) ref('a') $= a$
ref('b') $= b$

The semantic value of a predicate is the extension of the predicate. The extension of a one-place predicate 'F( )' is the collection of things that are F, and the extension of the two-place predicate 'R(,)' is the collection of ordered pairs such that the first member of the pair is R to the second member. So, for all x and all pairs $<x, y>$,

(4) $x \in$ ext('F( )')) $\equiv$ F(x)
$<x, y> \in$ ext('R(,)')) $\equiv$ R(x, y)

Next, we need a clause that tells us how the semantic values of 'a', 'b', 'F( )', and 'R(,)' contribute to the semantic value of 'F(a)' and 'R(a, b)':

(5) 'F(a)' is true $\equiv$ ref('a') $\in$ ext('F( )')
'R(a, b)' is true $\equiv$ $<$ref('a'), ref('b')$>$ $\in$ ext('R(,)')

From (3), (4), and (5) we can deduce (T_0).

Now to the semantics of temporally modified predications. The relational account moves from the semantics of atemporal predications to the semantics of temporal predications by building a time into the extension of the predicate: where the standard extension of 'F( )' is a class of ordinary objects, the extension of 'F(,)', according to the relational account, is a class of ordered pairs of ordinary objects and times. Here is how this kind of move can be avoided. As it is clear from the sentence-modifier account of the logical form of temporal predications, 'F( )' is a one-place predicate. The extension of this predicate is a class of ordinary objects. The predicate, however, does not have its extension simpliciter. The predicate has an extension only *relative to a time*. For example, relative to t_1 the predicate 'is happy' has as its extension the class containing only Charles and Suzie, whereas relative to t_2 the same predicate has as its extension the class containing only James and Zoe. This idea of temporally relativized extensions is standard in intensional semantics, where the idea is cashed out by saying that 'is happy' has as its *intension* a function from times to classes of ordinary objects. Given this apparatus, the move from the semantics of atemporal predications to the semantics of temporal predications can be informally described as building a time not into the extension of the predicate, but rather into the intension of the predicate. Note, however, that the notion of an intension need not be brought in for this proposal to work.

In order to implement this proposal, clause (3) remains in play but clauses (4) and (5) need revision. The semantic value of a predicate relative to a time t is the extension of the predicate relative to t. The extension relative to t of 'F( )' is the class of things that are F at t; similarly for 'R(,)'. Thus, for all x and all pairs $<x, y>$;

(6) $x \in \text{ext}_t(\text{`F( )'}) \equiv \text{At } t[\text{F}(x)]$
$<x, y> \in \text{ext}_t(\text{`R(,)'}) \equiv \text{At } t[\text{R}(x, y)]$

The clauses that tell us how the semantic values of '*a*', '*b*', 'F( )', and 'R(,)' contribute to the semantic value of 'At t[F(a)]' and 'At t[R(a, b)]' then look as follows:

(7) 'At t[F(a)]' is true $\equiv$ ref('a') $\in \text{ext}_t$('F( )')
'At t[R(a, b)]' is true $\equiv$ <ref('a'), ref('b')> $\in \text{ext}_t$('R(,)')

From (3), (6), and (7) we can deduce (T_1). This semantic account of temporal predication will be called the *intensional account*.

The first consequence of the intensional account to be pointed out is that it satisfies the constraint inherited from the semantic problem of change that an account of temporal predication make sure that reports of change do not imply a contradiction. The change-report 'Zoe is happy at t_1 and sad at t_2' does not imply the contradictory 'Zoe is happy and sad' because the predicates 'is happy' and 'is sad' have extensions only relative to a time. While Zoe is a member of the extension of 'is happy' relative to t_1 and a member of the extension of 'is sad' relative to t_2, Zoe is not a member of the extensions of 'is happy' and 'is sad' simpliciter. Once the problem of change has been characterized as a semantic problem, the intensional account is sensible and hard to miss. In the traditional framework set by the problem of change construed as a metaphysical problem, however, the intensional account has remained unnoticed.

The intensional account differs from the relational account in staying closer to the surface form of temporal predications and therefore in being able to take seriously certain intuitions that flow from this surface form. The intensional account allows shapes and colours to be properties instantiated relative to times, whereas the relational account turns shapes and colours into relations to times. The intensional account thus avoids metaphysically extravagant entities, such as shape-relations, and takes seriously the variation-intuition that change in shape is having different shapes at different times. However, the intensional account purchases its logical convergence with surface form at the cost of complicating the semantics of temporal predications via the notion of a predicate's having an extension only relative to a time. This semantic move will now be shown to have counter-intuitive consequences.

The incompatibility claim and cross-temporal predication

According to the intensional account, the temporal operator 'At *t*' functions as a sentence modifier that combines with a sentence '*s*' to form a new sentence 'At *t*[*s*]'. Now recall the sentential operator 'SIMP_{d}', introduced in Chapter 1, which has the function of making explicit that the sentence it governs is not governed by any sentential temporal operators, and that the sentence it governs has temporally unrelativized truth conditions (henceforth I will drop the subscript 'd', standing for detenserism, from the SIMP-operator). 'SIMP[F(*a*)]' is to be read as '*a* is F simpliciter'. Note that there are other ways of defining 'SIMP', such as the following:

(Def 1) $\text{SIMP}[s] =_{\text{df}} \exists t(\text{At } t[s])$
(Def 2) $\text{SIMP}[s] =_{\text{df}} \forall t(\text{At } t[s])$

These are two ways of defining 'SIMP' as a complex sentential temporal operator. There is nothing that prevents one from defining 'SIMP' in this way. But it is important that neither of these meanings is the meaning of 'SIMP' intended here. Here 'SIMP' is used as forming temporally unmodified, or timeless, predications, and not as forming complex temporal predications.[9]

What is the logical relationship between a temporal predication of the form 'At *t*[F(*a*)]' and an atemporal predication of the form 'SIMP[F(*a*)]', if the intensional account of temporal predication is correct? Suppose that

[9] Notice, however, that to deny that 'SIMP' is defined in the way suggested by (Def 1) or (Def 2) is not automatically to deny that 'SIMP[*s*]' is *equivalent* to '$\exists t(\text{At } t[s])$' or to '$\forall t(\text{At } t[s])$'. This is a different question, to be addressed in what follows.

'At *t*[F(*a*)]' is true. Assuming the intensional account, the extension of the predicate 'F( )', as it occurs in 'At *t*[F(*a*)]', is determined relative to the time mentioned by the temporal operator 'At *t*' that governs the statement of which the predicate is a constituent. That the predicate only has an extension relative to a time has the consequence that, if a statement in which the predicate occurs does not specify a time, then the predicate, as it occurs in this statement, lacks an extension, no matter what the subject of the statement is. That is, if the statement does not contain any temporal operators, then there is no time relative to which the extension of the predicate can be determined, and hence there is no extension. The point may be put in terms of intensions: the intension of 'F( )' is a function *f* from times to extensions. If a statement in which 'F( )' occurs is not governed by any temporal operators, then the statement fails to supply an argument for *f*, and hence *f* fails to supply an extension for the predicate. But, if there is no extension, then the statement cannot be true. In short: assuming the intensional account, if something is F at a time, then 'is F' has an extension only relative to a time, and hence '*x* is F simpliciter' cannot be true for any *x* because no time is specified to yield an extension. This may be called the *incompatibility claim* and stated perspicuously as follows:

(INC) $(\forall x)(\exists t(\text{At } t[\text{F}(x)]) \supset (\forall y)(\neg\text{SIMP}[\text{F}(y)]))$

This incompatibility is the important consequence of the intensional account that I will now show to be unacceptable.

The incompatibility claim (INC) faces an immediate counter-example if it is possible for a concrete object to share a property with an abstract object. Suppose that a piece of land and a geometrical figure can be exactly alike in shape by sharing the property of triangularity. The piece of land can be triangular at *t* and the geometrical figure can be triangular simpliciter: At *t*[Triangular(*L*)] & SIMP[Triangular(*F*)]. Given (INC), however, if the piece of land is triangular at a time, then nothing is triangular simpliciter, and hence the piece of land and the geometrical figure cannot have the same shape. This counter-example should not be given too much weight, since even hard-core Platonists may be sceptical about the supposition that abstracta can have shapes.

Further counter-examples involve only concrete things. These are counter-examples to the following weaker incompatibility claim that is entailed by (INC): anything is such that, if it is F at a time, then *it* is not F simpliciter. Or formally:

(INC*) $(\forall x)(\exists t(\text{At } t[\text{F}(x)]) \supset \neg\text{SIMP}[\text{F}(x)])$

In standard cases of temporal predication, such as 'Zoe is happy at t', we talk about an object or objects as they are at a particular time. There is an important range of cases, however, in which we talk about an object or objects as they are over a certain period of time, in which we trace an object or objects through time. In order to do this, we need to take a perspective on concrete matters from outside any particular time—we need to look at things *sub specie aeternitatis*. I shall call these cases *cross-temporal predications*. Some predications of this kind cause trouble for (INC*), and hence for the intensional account.

Here is an example. Suppose that, at t, Zoe exists and is identical to herself:

(8) At t[E(Zoe)]
At t[Zoe = Zoe]

Now consider the following sentence:

(9) Zoe, and only Zoe, is happy at t_1 and sad at t_2.

Notice first that (9) is not saying that only Zoe is happy at t_1 and only Zoe is sad at t_2. (9) is rather saying that Zoe is the only one who is both happy at t_1 and sad at t_2. So (9) is a cross-temporal predication, a case of tracking a particular individual through time. (9) must therefore be read as 'Zoe, and only Zoe, is such that at t_1(she is happy) and at t_2(she is sad)'. Or formally:

(10) At t_1[Happy(Zoe)] & At t_2[Sad(Zoe)] & SIMP[$\forall x$(At t_1 [Happy(x)] & At t_2[Sad(x)] $\supset x =$ Zoe)]

Here the universal quantifier and the identity statement '$x =$ Zoe' lie outside the scope of any temporal operators, as is indicated by 'SIMP'. Letting x be Zoe, and given that x is in the domain of the temporally unrestricted universal quantifier that ranges over all the things that exist simpliciter, (10) implies

(11) SIMP[E(Zoe)]
SIMP[Zoe = Zoe]

By (INC*), (8) and (11) are incompatible. Hence we have counter-examples to (INC*) involving the existence-predicate and the identity-predicate. This case shows that cross-temporal predication requires the atemporal predication of existence and identity, in addition to the standard temporal predication of existence and identity.

As an attempt to overcome the difficulty with (9), one might suggest that (9) contains an implicit temporal modifier 'At t_0' yielding:

(12) At t_0[Zoe, and only Zoe, is happy at t_1 and sad at t_2]

If the statement modified by 'At t_0' is unpacked along the lines of (10), then '$x =$ Zoe' and the universal quantifier lie within the scope of 'At t_0' and 'SIMP' disappears. Accordingly, (12) does not imply (11) but only the unproblematic

(13) At t_0[E(Zoe)]
At t_0[Zoe = Zoe]

There are two problems with this reply. First, the postulation of the implicit modifier 'At t_0' is *ad hoc*. The intended meaning of the original statement (9) is fully captured by (10). So what is the point of detecting a further temporal operator? What function does the new operator perform? Secondly, the postulation of a further temporal operator has no effect. Temporal predications such as 'At t_1[Happy(Zoe)]' end up inside the scope of a further temporal operator 'At t_0', and hence as multiply temporally modified. Think of the time specified by a temporal modifier as an answer to the question 'When does the fact that *a* is F obtain?' Then it is obvious that, once a time is specified, there is no further question to be answered by any other evaluation time; it is not sensible to ask 'When does the fact that *a* is F at *t* obtain?' Thus, a temporal modifier 'At *t*' that applies to a sentence that is already temporally modified has no effect; the operator 'At *t*' is vacuous.

This intuitive point can be backed up by reference to the semantics of temporal predication under discussion. According to the intensional account, a temporal operator 'At *t*' supplies the time relative to which the extension of the predicate is determined. Consider now the following multiply modified sentence:

(14) At t_0[At t_1[F(*a*)]]

Here the extension of 'F( )' is determined relative to t_1, since 'F(*a*)' is governed by 'At t_1'. Since the extension of 'F( )' can be determined only once, the second operator 'At t_0' lacks a semantic function altogether, and hence is vacuous. Thus, (12) collapses into the problematic (10) and the counter-example remains in play.

These counter-examples rest on the natural assumption that a temporal predication of the form 'At *t*[E(*a*)]' and an atemporal predication of the form 'SIMP[E(*a*)]' contain the same, unambiguous existence-predicate, and that 'At $t[a = a]$' and 'SIMP$[a = a]$' contain the same, unambiguous identity-predicate. Another strategy of dealing with the counter-examples lies in denying the truth of these assumptions, and to distinguish between different existence- and identity-predicates, or between different meanings of the existence-predicate and the identity-predicate, claiming that the predicates occurring in (8) are distinct from the predicates occurring in (11), or that the predicates in (8) and (11) are the same but invoke different meanings:

(15) At $t[\mathrm{E}_{①}(a)]$ & SIMP$[\mathrm{E}_{②}(a)]$
At t[Zoe $=_{①}$ Zoe] & SIMP[Zoe $=_{②}$ Zoe]

These sentences clearly do not clash with (INC*). But we have not found a way out unless we have found a principled basis on which to draw the distinction between different existence- and identity-predicates, or between different meanings of these predicates. As regards existence, some philosophers distinguish between concrete or physical existence and abstract or mathematical existence. This elusive distinction is not a candidate for existence$_{①}$ and existence$_{②}$, since in (15) the two notions of existence apply to the same thing. As regards identity, some philosophers distinguish between different notions of identity under a sortal, such as being the same person or being the same table. This distinction is not a candidate for identity$_{①}$ and identity$_{②}$, since in (15) the two notions of identity apply to the same kind of thing.

The only other approach I can see is to distinguish between a temporal and an atemporal notion of existence and identity. Focus on existence. The obvious candidate for the temporal notion of existence is existence at a time, whereas the obvious candidate for the atemporal notion of existence is existence simpliciter. The intensionalist may say that the existence-predicate, 'E( )', is ambiguous in that it has both a temporally sensitive intension, which is a function from times to extensions and corresponds to existence at a time, and a temporally insensitive extension, which corresponds to existence simpliciter. The problem with this proposal is that the distinction between existence simpliciter and existence at a time rests not on different meanings of the simple existence-predicate, but rather on different ways of modifying this predicate. The predicate 'λx(SIMP[E(x)])( )' corresponds to the notion of existence simpliciter, which is the obvious candidate for the atemporal notion of existence. The predicate 'λx(At t[E(x)])( )' corresponds to the notion of existence at a time, which is the obvious candidate for the temporal notion of existence. These two complex existence-predicates both contain the simple existence-predicate 'E( )' (or 'λx(E(x))( )'). Since the two complex existence-predicates already cover the distinction between atemporal and temporal existence, that distinction cannot form the basis for a further distinction between different senses of the simple existence-predicate 'E( )'—assuming that one notion of temporal existence and one notion of atemporal existence is enough. Thus, the distinction between a temporally sensitive intension and a temporally insensitive extension is not aligned with the distinction between existence at a time and existence simpliciter. These considerations apply to identity as well as existence. So it remains entirely unclear how the predicative differences in (15) should be justified.

To conclude, the counter-examples show that there is something seriously wrong with the incompatibility claims (INC) and (INC*). Since these claims are straightforward consequences of the intensional account of temporal predication, the cases also show that there is something seriously wrong with the intensional account. This criticism concerns cases of ordinary temporal predication that the intensional account fails to capture, and therefore is confined to the level of ordinary time. In the following section, I will look beyond ordinary time and consider the role of the intensional account in the explanation of temporal supervenience.

3.4 INTENSIONAL SUPERVENIENCE FAILURE

An account of temporal supervenience is a specification of the spacetime facts on which certain types of ordinary temporal fact supervene and an explanation of how they supervene. The ordinary temporal facts whose supervenience is to be explained here are facts of persistence and of temporal instantiation. In this section, I shall criticize the intensional account of temporal predication for its role in an account of temporal supervenience. I shall argue that the intensional account, just as the relational account, makes a plausible account of persistence supervenience and of temporal-instantiation supervenience unlikely.

Persistence supervenience

How objects are in ordinary time logically supervenes on how they are in spacetime—that is, the facts of persistence logically supervene on facts about spatiotemporal location. The main accounts of the spatiotemporal location of ordinary objects, each specifying a possible supervenience base of persistence, are three-dimensionalism and four-dimensionalism. An endurantist account of persistence supervenience attempts to explain how persistence supervenes on three-dimensionalist facts about spatiotemporal location, and a perdurantist account of persistence supervenience attempts to explain how persistence supervenes on four-dimensionalist facts about spatiotemporal location. Our question is whether any of these accounts of persistence supervenience is possible if the intensional account of temporal predication is correct.

Let us begin with three-dimensionalism. How can the intensionalist explain that the fact that an ordinary object occupies various instantaneous spacetime regions entails the fact that this object persists, that it exists at various times? The intensional account of temporal predication says that a report of persistence 'a exists at t_1 and a exists at t_2' is to be regimented as 'At t_1[E(a)] & At t_2[E(a)]', where 'At t_1' and 'At t_2' are primitive sentence modifiers that can be read as 'relative to t_1' and 'relative to t_2'. The regimented report of persistence is then assigned the truth conditions stated in (T_1). This logico-semantic picture of persistence does little to explain the link between an object a's spatiotemporal location and a's persistence. The intensional account thus needs to be combined with bridge principles if it is to be able to explain persistence supervenience.

The familiar construction of instants, such as t_1 and t_2, as maximal sets of simultaneous spacetime regions is not sufficient to bridge this conceptual gap. For, unless the intensionalist says more about what existing relative to a set of regions amounts to, it seems possible for a to occupy various instantaneous spacetime regions in accordance with three-dimensionalism without existing relative to any sets of instantaneous spacetime regions. Nor is the intensionalist able to bridge the gap by identifying existence with occupation, because existence is a property, and occupation is a relation. This difficulty may be avoided by identifying existence with the complex property of occupying some region. The proposal, more precisely, is that the predicate 'E( )' and the complex predicate '$\lambda x[\exists R(\mathrm{O}(x, R))]$( )' are associated with the same property. This move ties existence down to occupation, but it still falls short of explaining why occupying a region that is a member of t entails existing relative to t. The move explains only why occupying a region entails existing. What is worse, existence is a property instantiated relative to a time, while occupying a region is a property instantiated simpliciter. Thus, if existing is occupying a region, then a exists at t iff a occupies a region at t. Since a also occupies a region simpliciter, we have a further counter-example to the incompatibility claim, (INC*), which is a consequence of the intensional account.

The only strategy to bridge the gap that seems to be available to the intensionalist is the strategy of defining the meaning of 'At t[E(a)]' in terms of the occupation predicate in the following way:

(L4) At t[E(a)] $=_{\mathrm{df}} \exists R(R \in t \;\&\; \mathrm{O}(a, R)$

Informally, that *a* exists relative to *t* means that *a* occupies some member-region of *t*. (L4) gives the intensionalist the link that she requires. Assuming that *a* occupies various instantaneous spacetime regions—that is, assuming three-dimensionalism—and given that each of these regions is a member of a maximal set of instantaneous regions, the latter being times, (L4) entails that *a* exists relative to various times. Hence, three-dimensionalism entails persistence in the intensionalist's sense.

The first problem with definition (L4) is that it is incompatible with the intensional account. The intensionalist views 'At *t*' as a semantically primitive sentential operator. 'At *t*[E(*a*)]' is therefore an irreducibly temporally modified predication, and as such cannot be synonymous with an atemporal predication. The right-hand side of (L4), however, is an atemporal predication. If (L4) is adopted, then the temporal modifier is semantically reducible, which conflicts with the core of the intensional account. Even if, on the other hand, the temporal modification were allowed a reduction, (L4) would still be too strong for the same reason adduced against (L1), the three-dimensionalist meaning postulate discussed in the context of the relational account. (L4), like (L1), is too strong, in that it mixes elements from what appear to be fundamentally different conceptual schemes, elements that had better be kept apart. In order to avoid reductionism, the intensionalist might offer a weaker principle:

(L4*) $\Box$ [At t[E(a)] $\equiv \exists R(R \in t$ & O(a, R))]

For the reasons given in 3.2 with respect to (L1*) and (L2*), this equivalence is of no help in explaining supervenience. The problem, in short, is that (L4*) is a statement of what needs to be explained, rather than an explanation. As pointed out before, the most promising way of bridging the gap between occupation facts and persistence facts is through a semantic analysis of 'At *t*'. Since the intensionalist views 'At *t*' as semantically primitive, she cannot avail herself of this strategy. We are left with serious doubts concerning the availability of an endurantist account of persistence supervenience on the basis of the intensional account of temporal predication.

As regards a perdurantist account of persistence supervenience or an account with some mixture of three-dimensionalism and four-dimensionalism as supervenience base, analogous doubts arise concerning the possibility of any such account for the intensionalist. For the intensionalist still has the problem that existence cannot be identified with occupation, and she is still forced into the inadequate strategy of bridging the gap between occupation facts and existence facts by defining the meaning of an ordinary statement of temporal existence in terms of the concept of spatiotemporal occupation. The result is that the intensionalist lacks an illuminating explanation of persistence supervenience.

Temporal-instantiation supervenience

The problem of temporal-instantiation supervenience is to determine on which spacetime facts ordinary temporal-instantiation facts logically supervene and to explain how they supervene. Recall that the relationalist construes the ordinary facts of temporal instantiation as temporally unmodified, relational facts: having a property at a time is really bearing a relation to a time simpliciter. According to the intensional account of temporal predication, the ordinary facts of temporal instantiation are irreducibly temporally modified; having a property relative to a time cannot be further explained. Which spacetime facts underlie these temporally relativized facts?

Unlike the relationalist, the intensionalist can account for the instantiation of shape across spacetime in terms of monadic shapes, since the intensionalist also admits those monadic shapes at the level of ordinary time. What the intensionalist cannot do, however, is construe spatiotemporal instantiation of shape in terms of having a shape simpliciter, since having a shape relative to a time is incompatible with having a shape simpliciter, by the incompatibility claim (INC). If *a* is bent-shaped at *t*, then nothing is bent-shaped simpliciter, since the predicate 'is bent-shaped' has an extension only relative to a time. Thus, nothing in spacetime is bent-shaped simpliciter, given that the shape-predicates employed in spacetime discourse are the same shape-predicates employed in ordinary temporal discourse. Against the claim that distinct predicates are employed, or distinct meanings of the same predicate invoked, I argued in the previous section that there is no principled basis for such a distinction. Further, since the concept of a time is not part of the spacetime scheme, the intensionalist's account of ordinary temporal instantiation, according to which shapes are instantiated by objects relative to a time, is not also an account of spatiotemporal instantiation. In order to give an account of spatiotemporal instantiation, the intensionalist must first acknowledge the familiar bridge principle that identifies the times of ordinary time with certain entities of spacetime, and then relativize the instantiation of shapes in spacetime to these entities. Given that times are maximal sets of simultaneous spacetime regions, the intensionalist's account of spatiotemporal instantiation says that shapes are instantiated by objects relative to maximal sets of simultaneous spacetime regions. Accordingly, the intensionalist's facts of spatiotemporal instantiation entail her ordinary facts of temporal instantiation, which explains temporal-instantiation supervenience.

The intensionalist's account of spatiotemporal instantiation further answers the metaphysical problem of change. An object *a*'s having incompatible shapes at different times cannot be grounded in *a*'s having incompatible shapes simpliciter. So the metaphysical problem of change poses the task of avoiding a threat of contradiction. The intensionalist avoids contradiction by saying that shapes are distributed across spacetime in virtue of being instantiated relative to hyperplanes. An object *a*'s having incompatible shapes at different times is thus grounded in *a*'s having incompatible shapes relative to different hyperplanes.

The intensional account of temporal-instantiation supervenience and the relational account are structurally similar, and so are the doubts concerning their success. First, the intensional account, like the relational account, seems to reverse the order of explanation, in that the intensional account of spatiotemporal instantiation, which is meant to be an explanatory basis of ordinary temporal instantiation, looks more like a mere projection from the intensional account of ordinary temporal instantiation. Owing to the baroque complexity of the notion of having a shape relative to a set of simultaneous spacetime regions, this notion appears to be derived from the notion of having a shape relative to a time. But it would be wrong to let what goes on in ordinary time determine what goes on in spacetime. Spacetime comes first.

A second objection to the intensional account of temporal-instantiation supervenience results from the temporal-existence requirement (introduced in Section 3.2). This is the requirement that something can have a property at a time only if it exists at a time. Since the logical relations of ordinary instantiation facts and existence facts supervene on the logical relations of the spacetime facts that form their respective supervenience base, the spacetime fact that entails that *a* is F at a time should imply that *a* is spatiotemporally located, given the plausible assumption that ordinary facts of temporal existence supervene on facts of spatiotemporal location. The problem for the intensional account of spatiotemporal instantiation, as well as for the relational account, is that the intensional account fails to satisfy the temporal-existence requirement or, if it does satisfy the requirement, fails to explain why it satisfies the requirement. It seems possible for an object to have a shape relative to a set of simultaneous spacetime regions without occupying any portion of spacetime. This seems possible because the notion of having a shape relative to a hyperplane, where the relativization remains unexplained, seems to be completely independent from the notion of occupying a spacetime region simpliciter. Consequently, if there were a necessary connection between the intensionalist's notion of having a shape relative to a hyperplane and the notion of occupying a spacetime region simpliciter, then the intensionalist would be forced to admit this connection as a brute, mysterious fact.

The considerations in this section raise serious doubts as to the availability of a plausible intensionalist explanation of temporal supervenience. As in the case of the relational account, the failure to allow for such an explanation constitutes a significant flaw of the intensional account of temporal predication.

3.5 ADVERBIALISM

The semantic problem of change is essentially the problem of temporal predication (as was argued in Section 1.4). As accounts of temporal predication, the relational account and the intensional account are 'solutions' to the semantic problem of change. I shall conclude this chapter with a discussion of what is allegedly a further solution to the semantic problem of change: *adverbialism*. The question that will guide the discussion is whether adverbialism constitutes a further account of temporal predication.

In order to introduce the idea of adverbialism, it is necessary to return to the metaphysical gloss of the problem of change with which we started: to explain how change is possible, to explain how things can have different, incompatible properties at different times. The adverbialist solution to this problem is usually put by saying that the having, or instantiation, of incompatible properties by the same thing is temporally modified. In the traditional framework set by the problem of change, adverbialism has become a tag for an ontologically innocent solution, for a solution to the problem of change that avoids metaphysical extravagance as found in the view that apparent properties like shapes and colours are really relations between objects and times—a consequence of the relational account of temporal predication—and the view that shapes and colours are properties not of ordinary objects but of their temporal parts—a view that will be examined in Chapter 4.[10]

As it stands, the intuitive idea behind adverbialism—that the having of a property by an object is temporally modified—is underspecified. For the idea seems to make a trivial point that everyone can accept. Everyone agrees that the record is black just in case it has the property of being black (whether or not the right-hand side of this biconditional is read in an ontologically serious way). Likewise, everyone is willing to accept the temporally modified version of this biconditional: the record is black at *t* just in case it has the property of being black at *t*. But then there is a trivial sense in which having a property, such as blackness, is temporally modified: it is simply the pre-theoretic idea of having a property at a time. Relatedly,

[10] Recent proponents of adverbialism include Johnston (1987), Lowe (1987, 1988, 2000), Haslanger (1989*a*), van Inwagen (1990), and Rea (1998). Most of these authors say little or no more about their view than what has been said in the preceding paragraph. Charles (2000: app. 2) traces adverbialism and its modal analogue back to Aristotle.

the term 'adverbialism' suggests the view that the modifier in an ordinary temporal predication '*a* is F at *t*' is to be treated as a temporal adverbial. But that the modifier in a temporal predication functions, at the surface level, as an adverbial is again something everyone can accept. Moreover, the trivial observation that the surface form of a temporal predication 'Zoe is happy at *t*' contains a one-place predicate 'is happy' and a temporal adverbial 'at *t*' is the linguistic counterpart of the ordinary intuition that the statement says that Zoe has the property of being happy at time *t*. Construed in this way, adverbialism explains nothing, and hence fails to qualify as an account of temporal predication and change.

Progress can be made, though. The first non-trivial version of adverbialism is the intensional account of temporal predication. This fact, together with the intensional account itself, has remained largely unnoticed. The intensional account can be viewed as a version of adverbialism in so far as it is ontologically innocent: it avoids invoking any ontologically extravagant entities such as shape-relations or temporal parts of persons. A second non-trivial version of adverbialism has recently been discussed by David Lewis. Here is what Lewis thinks adverbialism, or *tensing the copula*, consists in:

> *Having* was originally thought to be a dyadic relation of things to properties; now it will instead be a triadic relation of things to properties and times. If you have at *t* the property *bent*, the property *bent* is unscathed: it is still the same old monadic intrinsic property we always thought it was. It is not replaced either by a relation or by a relational property.[11]

What connects 'tensing the copula' to the original, trivial kind of adverbialism is the intention to avoid 'tensing the predicate'—that is, turning shapes and colours into shape-relations and colour-relations—and to avoid 'tensing the subject'—that is, bringing in temporal parts of objects as the bearers of shapes and colours. Lewis directs a number of critical points against the proposal. These points will not be discussed here. My aim is not to defend adverbialism against Lewis's criticism. My aim is rather to deepen our understanding of adverbialism and of where it goes wrong. As far as discussion of Lewis-style adverbialism goes, I will rest content with raising a problem that is different and more serious than Lewis's.

Lewis-style adverbialism is meant to be a solution to the problem of change. When introducing the trivial version of adverbialism, we presupposed the original, imprecise construal of the problem of change. Lewis-style adverbialism, however, is more ambitious, and so, for it to be worth anything, it must qualify as an account of the problem of change properly

[11] (Lewis 2002: 5).

understood—that is, as an account of temporal predication. I will now show that Lewis-style adverbialism does not pass this requirement.

It will be helpful to begin by considering the following necessarily equivalent, unregimented temporal predications:

(16) *a* is F at *t*.
(17) *a* instantiates F-ness at *t*.

I shall assume, what is perfectly common, that we have here two different sentences—that is, that neither of these sentences reduces to, or is elliptical for, the other. This is an assumption about the form of (16) and (17), not about their content. As regards content, a realist about properties and relations might read (17) in an ontologically serious way as involving genuine reference to properties and relations. A Quinean nominalist, on the other hand, might read (17) as no more than a long-winded way of saying the same thing as (16). This controversy may be set aside. For the argument to follow is concerned primarily with questions about the form of (16) and (17) on which the realism–nominalism debate has no impact.

Lewis-style adverbialism says that an object *a* instantiates a property F-ness at a time *t* in virtue of bearing a *triadic* instantiation relation to F-ness and *t*. This thesis may be stated more precisely as the thesis that temporal predications with the surface form of (17) have the following logical form:

(17*) I(*a*, F-ness, *t*)

where 'I(, ,)' is a three-place instantiation-predicate ascribing, as Lewis says, a triadic instantiation relation to *a*, F-ness, and *t*. So far we have a proposal of how to treat the temporal modifier 'at *t*' on the surface of (17). This proposal is similar to the relational account in that it countenances no temporal modifier at the level of logical form, but avoids the unwelcome consequence that F-ness becomes an F-relation. However, we do not yet have a complete account of temporal predication. For a complete account would have to be an account of 'at *t*' not only in (17), but also in (16). Thus, the question is how to extend the logical thesis behind Lewis-style adverbialism to (16). Moreover, since (16) and (17) are equivalent, the task is to extend the logical proposal in such a way that this surface equivalence is preserved.

Finding a logical form of (16) that fulfils this task is hard. On the face of it, the possible logical functions of 'at *t*' in '*a* is F at *t*' are the following: (i) as sentence modifier; (ii) as predicate modifier; (iii) as subject modifier; and (iv) as eliminable. But none of these options leads to a temporal predication that is equivalent to (17*).

Option (i) is implemented by regimenting '*a* is F at *t*' as 'At *t*[F(*a*)]'. The sentence modifier 'At *t*' may be given the intuitive reading 'it is the case at *t* that'. Now consider the sentence 'It is the case at *t* that Zoe is happy'. Obviously, this sentence is equivalent to 'It is the case at *t* that Zoe has the property of being happy (or happiness)'. Formally, 'At *t*[F(*a*)]' is equivalent to 'At *t*[I(F-ness, *a*)]'. But 'At *t*[I(F-ness, *a*)]' contains a two-place instantiation-predicate, whereas (17*) contains a three-place instantiation-predicate. So 'At *t*[F(*a*)]' is not equivalent to (17*), and hence not the desired logical form of (16).

Option (ii) is to construe 'at *t*' as a predicate modifier that attaches to an *n*-place predicate to form a more complex *n*-place predicate: 'At *t*[F](*a*)'. Again, however, the result is not equivalent to (17*), which contains a three-place instantiation-predicate, but rather to 'I(*a*, F-ness-at-*t*)' which contains only a two-place instantiation-predicate, where F-ness-at-*t* is the complex property associated with the complex predicate 'At *t*[F]( )'.

Following option (iii) leads to 'F(*a*-at-*t*)', which is equivalent to 'I(*a*-at-*t*, F-ness)'. As in cases (i) and (ii), 'I(*a*-at-*t*, F-ness)' contains a two-place instantiation-predicate, whereas (17*) contains a three-place instantiation-predicate. So 'F(*a*-at-*t*)' is not the right logical form of (16) either. (I shall return to this account of the logical form of '*a* is F at *t*' in Chapter 4.)

The final option, option (iv), is to specify a logical form in which no temporal operator occurs—that is, to eliminate 'at *t*'. This option sounds most promising, since Lewis-style adverbialism proposes to get rid of 'at *t*' in the case of (17). The problem is that the only way of getting rid of 'at *t*' in '*a* is F at *t*' seems to be the relational account discussed earlier, which yields 'F(*a*, *t*)'. But this logical form is equivalent to 'I(*a*, F-relation, *t*)'. While the latter sentence contains a three-place instantiation-predicate, it also mentions an F-relation, whereas Lewis-style adverbialism mentions an F-property. In short, option (iv) of dealing with 'at *t*' in (16) leads to the relational account that Lewis-style adverbialism was designed to avoid.

So options (i)–(iv) each fail to provide the desired logical form of (16). Have we missed an option? Mark Johnston, who was among the first contemporary proponents of adverbialism, suggests reading '*a* is F at *t*' as '*a* is *t*-ly F'.[12] I can see three relevant ways of understanding this proposal. Johnston introduces the temporal adverb '*t*-ly' by analogy with such modal adverbs as 'actually' and 'possibly' in '*a* is actually F' and '*a* is possibly F'. It is standard to view these modal adverbs as functioning as sentence modifiers yielding 'Actually/possibly[F(*a*)]' or as predicate modifiers yielding 'Actually/possibly[F](*a*)'. By analogy, the suggestion is

[12] Johnston (1987: 127–8).

to treat 'at *t*' as a sentence modifier or as a predicate modifier. These are options (i) and (ii), which have already been discussed and rejected.

One might, however, develop Johnston's idea in a different way, namely to understand Lewis's slogan 'tensing the copula' literally, and to claim that 'at *t*' logically functions neither as a sentence modifier nor as a predicate modifier nor as a subject modifier, but rather as a modifier of the copula 'is', yielding '*a* is-at-*t* F'.[13] This proposed reading of (16) may be formalized as follows:

(18) At *t*[is]F(*a*)

The problem with this strategy of copula modification lies in giving a semantics for (18). Informally speaking, the account will look like this: *a* is-at-*t* F iff *a* is a member of the extension of 'is-at-*t* F'. How is the extension of this complex predicate determined? The extension of the predicate 'is-at-*t* F' must be determined compositionally on the basis of the semantic values of 'is-at-*t*' and 'F'. Furthermore, the semantic value of 'is-at-*t*' must be determined on the basis of the semantic values of 'is' and '*t*'. That means that the copula 'is' is required to have semantic significance. A straightforward way of giving the copula and the adjective 'F' semantic significance is to claim that 'is' is synonymous with 'instantiates' and 'F' is synonymous with 'F-ness'. The statement '*a* is-at-*t* F' would then be trivially equivalent with '*a* instantiates-at-*t* F-ness'. Formally, (18) would be equivalent to:

(19) At *t*[I](*a*, F-ness)

This move, however, is highly implausible, given the fundamental difference in syntactic function between the copula 'is' and the predicate 'instantiates' and between the adjective 'F' and the singular term 'F-ness'. For how can expressions with different syntactic functions have the same semantic value? Moreover, even if (18) were equivalent to (19), this would not further the adverbialist's project. For (19) contains a two-place instantiation predicate modified by 'At *t*', whereas (17*) contains an unmodified three-place instantiation predicate. Therefore, (18) cannot be the logical form of (16) that the Lewis-style adverbialist was looking for.

In general, it is unclear how the suggestion that the copula is modified should be worked out into a proper semantic account of temporal predications and change-reports. The copula is most naturally construed as a semantically insignificant part of the predicate 'is F'. The semantic value of the predicate 'is F' is not a function of the semantic values of 'is' and 'F',

[13] This strategy is adopted by Jonathan Lowe in his (2002: 47–9).

since 'is' and 'F' lack a semantic value altogether. All of this is post-Fregean orthodoxy.

To conclude, options (i)–(iv) plus copula-modification seem to exhaust the possible logical functions of 'at *t*' in '*a* is F at *t*'. Since none of these options generates a plausible logical form of (16) that is equivalent with the adverbialist's (17*), we have reason to believe that there is no such logical form. Hence, Lewis-style adverbialism does not deliver a coherent account of temporal predication. To be precise, Lewis-style adverbialism does not merely fail to provide a plausible account of temporal predication; it even fails to qualify as such an account. Lewis-style adverbialism is therefore a mirage.

4

Four-Dimensionalist Supervenience

Now that we have considered several failed attempts to explain temporal supervenience, it is time to examine an approach that works. In the opening section of this chapter, we will meet the four-dimensionalists with their theory of temporal parts, a significantly different account of temporal predication from the ones considered in the previous chapter, and the four-dimensionalists' impressively elegant explanation of temporal supervenience. It will further be shown that our questions of temporal supervenience as well as the four-dimensionalists' answers to these questions have interesting spatial and modal analogues. Despite the four-dimensionalists' success in explaining the supervenience of facts of persistence and change, the news in subsequent sections will be that not all is well in the four-dimensionalist universe.

4.1 THE TEMPORAL-PARTS ACCOUNT

Since the problem of temporal supervenience rests on the semantic problem of temporal predication and the metaphysical problems of spatiotemporal location and spatiotemporal instantiation (see Chapter 1), we need an account of temporal predication as well as of spatiotemporal location and instantiation before we can proceed to an account of temporal supervenience. In this section, I shall construct four-dimensionalist accounts of these three types.

The representational account of temporal predication

An account of temporal predication is a specification of the logical and semantic function of the temporal modifier 'at *t*' in temporal predications with the surface form '*a* is F at *t*'. Such an account can take various forms. The temporal predication '*a* is F at *t*' may be given a logical form in which 'at *t*' has disappeared. Or '*a* is F at *t*' may be given a logical form that still contains a temporal modifier. If there is a temporal modifier at the level of logical form, there is a choice of construing this modifier as semantically reducible or as semantically irreducible. In the relational account of temporal predication we have seen an account of '*a* is F at *t*' that gets rid of 'at *t*'. In the intensional account we have seen an account of '*a* is F at *t*' that admits a temporal modifier at the level of logical form and construes this modifier as semantically irreducible. Let us now consider the remaining option of detecting a temporal modifier in the logical form of '*a* is F at *t*', but of viewing this modifier as semantically reducible.

We start with the sentence-modifier reading of 'at *t*' in '*a* is F at *t*', which was also employed by the intensional account. According to this reading, the logical form of '*a* is F at *t*' contains a one-place predicate and a sentential temporal operator 'At *t*', which attaches to a sentence to form a new sentence, yielding: At $t[F(a)]$. This logical form stays close to the surface form of '*a* is F at *t*'. So far, so familiar.

To say that 'At *t*' is semantically reducible means that 'At $t[F(a)]$' has truth conditions in which the modifier 'At *t*' does not occur. This reducibility may be achieved in the following way. First, consider another ordinary sentence modifier, the modifier occurring in such sentences as 'Zoe is happy according to the *Guardian*'. The sentence modifier 'according to the *Guardian*', or more generally, 'according to *x*' is a representational modifier and can be given the following definition:

(Def1) a is F according to $x =_{df} x$ represents a as being F.

This is a reductive definition of the modifier 'according to *x*' in virtue of specifying a statement that is synonymous with '*a* is F according to *x*' but does not contain 'according to *x*'. This reductive strategy may now be applied to 'At $t[F(a)]$' by formalizing 'according to *x*' as 'Ac *x*' and interpreting 'At *t*' as 'Ac *t*', to be read as 'according to *t*', yielding 'Ac $t[F(a)]$'. The representational temporal modifier 'Ac *t*' may then be defined along the lines sketched above:

(Def2) Ac $t[F(a)] =_{df} t$ represents a as being F.

So temporal predication may be interpreted as representational predication. Although intuitively well understood, present concerns make it reasonable to demand further clarification of what exactly is meant by saying that *x* represents *a* as being F. In particular, we want to know how the predicate 'is F' functions in this phrase. In response, representing *a* as being F may be further analysed in terms of the notion of a *representative*:

(Def3) x represents a as being F $=_{df}$ a has a representative in x that is F simpliciter.

If 'At $t[F(a)]$' is interpreted as 'Ac $t[F(a)]$', and if (Def2) and (Def3) are put together, then the right-hand side of (Def3) with '*x*' replaced by '*t*' gives the meaning of 'At t[F(a)]'. Since the right-hand side of (Def3) is a temporally unmodified, or atemporal, predication, monadic temporal predications may be given temporally unmodified truth conditions:

(T_2) 'At t[F(a)]' is true $\equiv \exists x(\text{Rep}(x, a, t) \,\&\, \text{F}(x))$

where 'Rep(x, a, t)' is to be read as 'x is a's representative in t'. Following analogous steps, a similar semantic reduction may be achieved for dyadic temporal predications:

(T'_2) 'At t[R(a, b)]' is true $\equiv \exists x \exists y(\text{Rep}(x, a, t) \,\&\, \text{Rep}(y, b, t) \,\&\, \text{R}(x, y))$

This is the *representational account* of temporal predication.[1]

The purpose of the representational account is to reduce temporal predication to atemporal predication by turning temporal predication into a form of representational predication. But there the account stops. Its purpose is not to explain representation as well. In particular, (Def3) clarifies the role of the predicate 'is F' in representational claims without also telling us what a representative is, and hence how the representing is done. How representation works is an important but different question to be addressed shortly.

The representational account unpacks the meaning of temporal modifiers in representational terms. In Chapter 3, various postulates concerning the meaning of temporal predications were criticized for their reductionist tendency to define ordinary temporal concepts in terms of spacetime concepts. The definition of temporal operators as representational operators cannot be accused of conflating different conceptual schemes. For the representational account defines the meaning of temporal operators in terms of representational properties of times, not of spacetime regions, and hence stays firmly within the realm of ordinary time.

The representational account of temporal predication is an alternative to the relational account and the intensional account discussed in the previous chapter. These accounts have already been shown to be objectionable at the level of ordinary time, in addition to raising problems concerning temporal supervenience. Leaving supervenience issues for later, does the representational account fare better with respect to the mentioned difficulties? The relationalist is bound to view the variation-intuition along with our ordinary conception of change as insignificant and misleading, since the characteristic feature of the relational account is that it assigns temporal predications a logical form that diverges significantly from their surface form; according to the relationalist, it is not strictly and literally true that objects vary in their

[1] A similar strategy of interpreting modal predication as representational predication is adopted by Lewis (1986*a*: 194–6). See Sect. 4.2.

properties over time. Assuming the methodological stance of optimism, which attempts to take ordinary intuitions seriously, an account of logical form that stays close to the surface form of temporal predications is to be preferred to the relational account. By construing temporal modifiers as sentence modifiers, the representationalist is able to take seriously the variation-intuition and the ordinary conception of change.

The intensional account was shown to have a consequence with many counter-examples, the incompatibility claim (INC), which says that if something is F at *t*, then nothing is F simpliciter, and its weakened version (INC*) which says that if something is F at *t*, then it is not F simpliciter. The representational account has no such consequence, since being F at *t* is analysed in terms of being F simpliciter. In fact, this semantic reduction of temporal to atemporal predication is the point of the representational approach. Accordingly, the representational account accommodates counter-examples to the incompatibility claim with ease, such as the example that a person is self-identical at some time *t* and self-identical simpliciter, which is unproblematic because it just says that both a person and its representative in *t* are self-identical simpliciter.[2] Compared to the relational account and the intensional account, then, the representational account looks promising.

Four-dimensionalism, temporal parts, and spatiotemporal instantiation

Now that we have given an account of the temporal dimension of ordinary language, let us move on to the task of giving an account of the temporal dimension of ordinary objects—that is, the metaphysical problems of spatiotemporal location and spatiotemporal instantiation. The task is to specify a spatiotemporal supervenience base for ordinary facts of temporal existence and of temporal instantiation. In Chapter 2, we distinguished between the two main accounts of spatiotemporal location: three-dimensionalism and four-dimensionalism. In this chapter, we want to find out about the four-dimensionalist's prospects of explaining temporal supervenience. Four-dimensionalism is the thesis that an ordinary object occupies a single spacetime region that is temporally extended (see Section 2.1). Starting from here, how are ordinary properties, such as colours and shapes, instantiated

[2] The last example is based on cross-temporal predication. Although the examples drawn from cross-temporal predication pose no difficulty for the representational account, it will be shown in Sect. 4.5 that cross-temporal predications themselves do pose a serious problem for the temporal-parts account of temporal supervenience of which the representational account is a part.

in spacetime? This is the problem of spatiotemporal instantiation. There is more than one way for the four-dimensionalist to handle this issue. But there is an obviously best and most elegant way, making use of temporal parts of objects. In Section 2.2, I developed a theory of temporal parts as an extension of four-dimensionalism. This theory presupposes various definitions that I will not repeat here: the definitions of the notions of an instantaneous and an extended temporal part of a spacetime region, (ITP:R) and (ETP:R), and the definitions of the corresponding notions of an instantaneous and extended temporal part of an object, (ITP:O) and (ETP:O). The theory was then stated in the following three theses:

(T1)/(4D) (i) an ordinary object occupies a unique spacetime region, and (ii) this spacetime region is temporally extended.

(T2) If an ordinary object x occupies a spacetime region R, then for every instantaneous and extended temporal part R' of R there is a part x' of x such that (i) x' occupies R', and (ii) x' does not occupy any other spacetime region.

(T3) An ordinary object is a sum of temporal parts that is maximal under some unity relation.

Given definitions (ITP:O) and (ETP:O), it follows from (T2), the doctrine of arbitrary temporal parts, that, if x occupies region R, then x has instantaneous and extended temporal parts that occupy corresponding instantaneous and extended temporal parts of R. Thus, it follows from (T1)/(4D) and (T2) that ordinary objects have instantaneous and extended temporal parts. In other words, it follows that ordinary objects are 'spacetime worms' with parts along the temporal dimension as well as the spatial dimensions. Thesis (T3) then explains how the temporal parts of a spacetime worm need to be related so that this worm counts as an ordinary object.

With the theory of temporal parts in place, it is only a short step to an account of how ordinary properties, such as shapes, are instantiated in spacetime. First remember that there are no times in spacetime, and hence that the ordinary notion of having a shape at a time has no application in the spacetime conception. Instead, shapes must be had simpliciter. But which component of occupied spacetime instantiates these properties simpliciter? The four-dimensionalist replies: temporal parts of ordinary objects. In spacetime, an ordinary object—a spacetime worm—has no ordinary shapes at all; rather, different ordinary shapes are had by its temporal parts. If an ordinary object did have shapes simpliciter, then there would be no hope of explaining the supervenience of ordinary facts of change in shape and in colour on spacetime facts. For x's having a bent shape at t_1 and x's having a straight shape at t_2 would be grounded in x's having a bent shape simpliciter and x's having a straight shape simpliciter, which is contradictory. By letting different temporal parts of ordinary objects have incompatible properties, the four-dimensionalist gives a coherent account of the spatiotemporal supervenience base of ordinary facts of change, and hence of the metaphysical problem of change.

We distinguished between instantaneous and extended temporal parts of an object. The temporal-parts account of spatiotemporal instantiation, as stated so far, says that in the spacetime conception ordinary properties are instantiated simpliciter by temporal parts of ordinary objects. Does this account hold for the object's instantaneous and extended temporal parts alike? The answer is 'no'. For reasons concerning temporal-instantiation supervenience to be given below, ordinary properties are instantiated only by instantaneous temporal parts of objects.

The temporal-parts account of the spatiotemporal supervenience base of ordinary facts of temporal instantiation raises a question of generality. It is clear that ordinary facts about the instantiation of temporary properties—properties that an object has at some times of its existence and lacks at others—logically supervene on the spatiotemporal location and properties of temporal parts. But how about permanent properties that an object instantiated at all times at which it exists? Could the four-dimensionalist say that ordinary facts about the instantiation of permanent properties logically supervene on the spatiotemporal location and properties of ordinary objects themselves, as opposed to their temporal parts? The reason why the four-dimensionalist should not adopt different accounts of the supervenience base of ordinary facts of temporal instantiation, depending on whether the properties are temporary or permanent, is that all these facts are expressed by the same kind of temporal predication: *a* is F at *t* (in the monadic case). As we will see in detail in the next subsection, an explanatory link between ordinary temporal facts of instantiation and spatiotemporal facts of instantiation rests on a semantic account of the modifier 'at *t*' in '*a* is F at *t*'. To specify different types of supervenience base for facts of temporal instantiation thus requires different semantic treatments of 'at *t*' depending on whether 'is F' is a temporary or a permanent predicate of *a*. A policy of caution suggests that such an ambiguity in 'at *t*' should not be posited unless we are really forced to do so, unless there are really compelling theoretical or intuitive reasons that suppose that such an ambiguity really exists.[3] In the case of 'at *t*', there is no such compelling reason. The postulation of an ambiguity in 'at *t*' is unjustified, because the question how 'at *t*' should be treated semantically is clearly independent of the question whether *a* is F at all times of its existence or only at some of those times. To say that a semantic treatment of 'at *t*' in '*a* is F at *t*' must take into account

[3] For this general 'policy of caution' concerning ambiguities, see Kripke (1977: 401).

at which times *a* exists is like saying that knowing what it means to say that Suzie is married requires knowing how many times Suzie was and will be married. Assuming, then, that the four-dimensionalist wants to give a unitary treatment of temporal modification, she is committed to the temporal-parts account of spatiotemporal instantiation in full generality.

Are there alternative accounts of spatiotemporal instantiation available to the four-dimensionalist? The temporal-parts account says that shapes are properties instantiated simpliciter, not relativized to anything. Alternatively, shapes could be construed as relations—an option familiar from the relational account—or shapes could be construed as properties the instantiation of which is somehow relativized—an option familiar from the intensional account. As we saw earlier, the unrelativized instantiation of shape-relations and the relativized instantiation of shape-properties are independent of any particular mode of spatiotemporal location: that an object has a shape-relation simpliciter or that an object has a shape-property relative to, say, a spacetime region or to a set of spacetime regions does not imply anything about how the object is located in spacetime, and indeed does not even imply that the object is spatiotemporally located at all. This has two consequences, a good one and a bad one.

The good consequence is that these accounts of spatiotemporal instantiation are compatible with any account of an object's spatiotemporal location, and hence that they are both compatible with four-dimensionalism. This compatibility is interesting because it makes available to the four-dimensionalist an account of spatiotemporal instantiation that does not require temporal parts. For an object may occupy a unique, temporally extended spacetime region, have no temporal parts corresponding to the temporal parts of this region, and yet have shape-relations to spacetime regions or sets of spacetime regions, or it may have shape-properties relative to spacetime regions or sets of spacetime regions. Of course, a four-dimensionalist might equally run these kinds of accounts of spatiotemporal instantiation *with* temporal parts saying that an object's temporal parts, as opposed to the object itself, have these shape-relations simpliciter or that an object's temporal parts have these shape-properties relative to some component of spacetime. In the light of the availability of the temporal-parts account of spatiotemporal instantiation, according to which shapes are properties instantiated simpliciter, the account employing shape-relations and the one employing relativized shape-properties appear needlessly complex. But there is a more substantial objection to these inelegant four-dimensionalist alternatives to the temporal-parts account.

The bad consequence is that the compatibility of unrelativized shape-relations and of relativized shape-properties with any account of spatiotemporal location goes against the temporal-existence requirement that says that an object can have a property at a time only if it exists at a time. Assuming that facts about ordinary time and their logical relations supervene on facts about spacetime and their logical relations, we get the requirement that the facts of spatiotemporal instantiation underlying ordinary facts of temporal instantiation imply certain facts of spatiotemporal location underlying ordinary facts of temporal existence (see Section 3.2). The temporal-existence requirement, in other words, places a constraint on how an account of spatiotemporal instantiation is to be related to an account of spatiotemporal location, a constraint that weighs heavily against combining four-dimensionalism with any of the accounts of spatiotemporal instantiation encountered in the context of the relational and the intensional account.

The temporal-parts account of spatiotemporal instantiation straightforwardly captures the temporal-existence requirement that *a* can have a property at a time only if *a* exists at time. The task is to specify a fact A of spatiotemporal instantiation that entails that *a* has a property at a time and a fact B of spatiotemporal location that entails that *a* exists at a time, such that A implies B. According to the temporal-parts account, the spacetime fact underlying *a*'s being F at a time *t* is the fact that *a* has a temporal part that is F and that occupies a spacetime region that is a member of *t*. Moreover, the spacetime fact underlying *a*'s existing at a time *t* is the fact that *a* has a temporal part that occupies a spacetime region that is a member of *t*. Since the fact that *a* has a temporal part that occupies a member-region of *t* and that is F trivially implies the fact that *a* has a temporal part that occupies a member-region of *t*, the temporal-existence requirement is satisfied.

These considerations suggest that the temporal-parts account of spatiotemporal instantiation is on the right track. But the four-dimensionalist is not yet in a position to celebrate. As I have made clear from the start, the main criterion of evaluating accounts of spatiotemporal location and instantiation as well as accounts of temporal predication is how well they play together in explaining temporal supervenience. Thus, the merits of the temporal-parts account of spatiotemporal instantiation cannot become fully apparent until we turn to temporal supervenience.

Putting the pieces together: The temporal-parts account of temporal supervenience

The four-dimensionalist puts forth two supervenience theses: (*a*) the facts of persistence logically supervene on facts about the spatiotemporal location of temporal parts of objects; (*b*) the facts of temporal instantiation logically supervene on facts about the atemporal instantiation of properties by temporal parts of objects. So far (*a*) and (*b*) are assertions of brute facts.

The remaining task is to reduce this bruteness, to explain persistence supervenience and temporal-instantiation supervenience.

In order to do so, let us return to the representational account of temporal predication. This account with its semantic interpretation of temporal modifiers as representational modifiers yields the following reductive equivalence:

(RED) $\Box[\text{At } t[\text{F}(a)] \equiv \exists x(\text{Rep}(x, a, t) \,\&\, \text{F}(x))]$

This principle construes representation of *a* by *t* in terms of a representative of *a* in *t*. What is a representative of *a* in *t*? It is obvious that *a*'s representative has to be distinct from *a*. For, if *a* is its own representative in *t* and is F, then the statement that *a* is F according to t_1 and not F according to t_2 entails that *a* is F and not F. So how does temporal representation work? Pursuing this question will lead us to an account of temporal supervenience.

Representation can function in a variety of different ways. As far as ordinary forms of representation go, there is, first, *linguistic* representation. A linguistic representative is a linguistic term—for instance, a name—that is hooked up to its bearer by convention. Linguistic representation works by saying. By containing a linguistic representative, a name, of Charlie, the statement 'Charlie is happy' says that Charlie is happy. Therefore the statement represents Charlie as being happy; Charlie is happy according to this statement. Another common kind of representation is *pictorial*. A pictorial representative may be construed as a picture in a generalized sense in which paintings, statues, and working models count as pictures. Pictorial representation works largely by qualitative similarity. A statue, by being overall more or less similar to the person of which it is a statue and by having an oddly shaped marble head, represents the person as having a disastrous haircut. Here the representative of the person, the statue, is also that which represents the person as being a certain way. But if the statue is part of a larger model, then the model represents the person as being a certain way by containing a representative that is distinct from the model.[4]

Four-dimensionalist temporal representation, as I view it, differs from both linguistic and pictorial representation. A temporal representative of an ordinary object *a* is a temporal part of *a*. For such a representative to be 'in' a time *t* is for the temporal part to occupy a spacetime region that is a member of the set of regions that is *t*:

($\text{REP}_{4\text{D}}$) $\Box[\text{Rep}(x, a, t) \equiv \text{TP}(x, a) \,\&\, \exists R(\text{O}(x, R) \,\&\, R \in t)]$

[4] This taxonomy is loosely based on Lewis's distinction between linguistic and pictorial ersatzism (Lewis 1986*a*: 2.1–2.3).

Thus, a time t represents a as being F by having as a member a spacetime region that is occupied by a temporal part of a that is F simpliciter. Note that (REP_{4D}) is not a thesis about what having a representative in t means, but rather a thesis about the essential nature of temporal representation. This form of representation is nothing like linguistic representation, but it resembles pictorial representation. Just as a model of a city and its inhabitants can represent a person as waving by containing a statue of the person that is waving, so a time can represent a person as being happy by 'containing' a temporal part of the person that is happy. We may add that, just as the statue itself, in addition to the model, also represents the person as waving, so the temporal part itself, in addition to the time, may be viewed as representing the person as happy. This case, however, in which the representative and that which does the representing are identical, is irrelevant in the present context and therefore does not require further attention.

Temporal supervenience can now be explained. The four-dimensionalist account of temporal representation functions as a bridge principle linking the four-dimensionalist's facts of spatiotemporal location and instantiation with the ordinary facts of temporal existence and instantiation via the representational account of temporal predication. Beginning with the sentence-modifier reading of 'a is F at t', 'At t[F(a)]', then interpreting 'At t' as representational, which yields (RED), and finally adding (REP_{4D}), delivers the following biconditional, which is the backbone of the four-dimensionalist account of temporal supervenience (I give the formal version for perspicuity and the informal version for later reference):

(TS_{4D}) $\Box[\text{At } t[\text{F}(a)] \equiv \exists x \exists R(\text{TP}(x, a) \,\&\, \text{O}(x, R) \,\&\, R \in t \,\&\, \text{F}(x))]$
Necessarily, a is F at t iff a has a temporal part that is F simpliciter and that occupies a spacetime region that is a member of t.

Now recall supervenience theses (a) and (b): the facts of persistence logically supervene on facts about the spatiotemporal location of temporal parts of objects; and the facts of temporal instantiation logically supervene on facts about the atemporal instantiation of properties by temporal parts of objects. These theses are explained by (TS_{4D}) in virtue of being consequences of this principle. That thesis (b) is a consequence of (TS_{4D}) is obvious. To see that thesis (a) is a also a consequence of (TS_{4D}), apply the latter to an ordinary existence statement 'a exists at t': a exists at t iff a has a temporal part that exists simpliciter and that occupies a member-region of t. This may be simplified: a exists at t iff a has a temporal part that occupies a member-region of t. Hence, facts about the spatiotemporal location of an object's temporal parts entail facts about the object's temporal existence, which is what thesis (a) asserts. The four-dimensionalist can thus explain persistence supervenience and temporal-instantiation supervenience by constructing (TS_{4D}) out of the representational account of temporal predication and the temporal-parts account of temporal representation.

Since the *t*s in '*a* is F at *t*' are instants—that is, sets of instantaneous regions—the only temporal parts of *a* that play a role in the explanation of temporal supervenience are instantaneous temporal parts. Earlier the question came up whether extended temporal parts should also be allowed to have ordinary properties simpliciter. Although it would not be inconsistent to let an extended temporal part made up of happy instantaneous temporal parts be happy as well, these properties of extended temporal parts play no theoretical role. They are redundant. So why postulate them?[5]

Let us take stock. The previous chapter concluded that the relational and intensional accounts of temporal predication do not seem to allow a plausible account of temporal supervenience. In this chapter, these accounts of temporal predication were rejected in favour of the representational account that allows a smooth account of temporal supervenience, the four-dimensionalist temporal-parts account. I shall close this section by considering two variants of the temporal-parts account that do not rest on temporal representation.

Short cuts: subject modifiers and ellipsis

Assume four-dimensionalism, the theory of temporal parts, and the temporal-parts account of spatiotemporal instantiation. The question is whether we can get to the temporal-parts account of temporal supervenience, (TS_{4D}), and to all its benefits without having to go via the representational account of temporal predication, that is without (RED) and (REP_{4D}). In other words, is there a short cut to (TS_{4D})?

The four-dimensionalist might contemplate two such short cuts. The first proposal is radically direct: (TS_{4D}) is true because its left-hand side is elliptical for its right-hand side—that is, its left-hand side when spelled out in full just is its right-hand side. Informally, '*a* is F at *t*' is short for '*a* has a temporal part that is F simpliciter and that occupies a member-region of *t*'. This proposal amounts to a logical elimination of 'at *t*' plus an account of temporal supervenience in one move. Let us call this the *ellipsis account* of temporal predication. The second proposal is to construe 'at *t*' as a subject modifier attaching directly to '*a*' yielding: F(At *t*[*a*]). The next step is semantically to reduce this subject modifier by interpreting the complex singular term 'At *t*[*a*]' as 'the temporal part of *a* that occupies a

[5] In Sect. 4.5, I will discuss an alleged reason for assigning extended temporal parts of ordinary objects a serious theoretical role in an account of temporal supervenience.

member-region of *t*'. This semantic reduction of 'At *t*' immediately yields (TS_{4D}). Let us call this the *subject-modifier account* of temporal predication. To emphasize, the temporal-parts account of temporal supervenience is a direct consequence of both the ellipsis account and the subject-modifier account. This makes the accounts simpler than the representational account, which needs, in addition to (RED), the bridge principle (REP_{4D}). But they come with negative baggage that quickly outweighs their simplicity. I shall make two objections.

The first objection is methodological. How are the ordinary conception of time and the spacetime conception related according to the ellipsis account and the subject-modifier account? Both accounts say, in different ways, that the ordinary conception is the spacetime conception in disguise. So each account is objectionably reductionist. By admitting an optimist stance, the representational account is favourable, avoiding a conflation of ordinary time and spacetime.

As reductionists, the friends of the ellipsis account and of the subject-modifier account claim an important advantage over a pessimist and an eliminativist stance: since the ordinary conception of time just is the spacetime conception in disguise, and since the spacetime conception describes the world correctly, the truth of the ordinary conception is preserved. My second objection to the ellipsis and subject-modifier accounts is that this is not entirely true. An important part of the ordinary way of viewing things is lost: our conception of change.

The no-change objection

Consider the following objection to the theory of temporal parts, known as the *no-change objection*.[6] It is perfectly commonplace to say that, when something changes in shape over time, it has different shapes at different times—it loses (or gains) its shapes. So our ordinary conception of change over time is temporal property-variation—having different, incompatible properties at different times. I earlier called this the 'variation-intuition'. The objection to the theory of temporal parts is that the theory and its account of the instantiation of shapes clash with the variation-intuition. For, if all shapes are instantiated by temporal parts of things, then there are no shapes left to be instantiated by ordinary things themselves. Since the shapes of temporal parts do not vary, nothing ever changes in shape in the ordinary sense. In short, the theory of temporal parts fails to capture the facts of change.

[6] Versions of this objection can be found in Lowe (1988), Haslanger (1989*b*), Hinchliff (1996), and Sider (2001: 5.2).

This objection has no force if the four-dimensionalist adopts the representational account of temporal predication. The objection does go through, however, if the four-dimensionalist adopts either the ellipsis account or the subject-modifier account. The variation-intuition is closely tied to the surface form of '*a* is F at *t*', which contains a one-place predicate and a sentential temporal adverbial 'at *t*', and which can therefore be expressed by saying that *a* has the property of being F at *t*. Thus, a logical form of '*a* is F at *t*' that stays close to its surface form is able to capture the variation-intuition, while a logical form that diverges from the surface is not. In the relational account of temporal predication we have already seen a case of an account that fails to capture the variation-intuition. Objects do not really have properties at times. Strictly speaking, objects bear relations to times. The ellipsis account and the subject-modifier account have a similar consequence. The former implies that an object does not really have a property at a time, and hence does not really change in the ordinary sense, because to say that an object has a property at a time is just short for saying that the object has a temporal part that has the property simpliciter. The subject-modifier account implies that an object does not really change in the ordinary sense, because to say that *a* has a property at *t* is really to say that *a*-at-*t* has the property simpliciter, where *a*-at-*t* is *a*'s temporal part at *t*. The representational account, on the other hand, stays close to the surface form of '*a* is F at *t*' by specifying a logical form that contains a one-place predicate and a sentence modifier, 'At *t*[F(*a*)]'. The representational account, therefore, allows that, strictly and literally speaking, an object has a property at a time, and hence that objects really change in the ordinary sense given by the variation-intuition. This is a further point favouring the representational account over the ellipsis account and the subject-modifier account.[7]

As I announced at the beginning of this chapter, despite the four-dimensionalist's success in explaining temporal supervenience, not all is well in the four-dimensionalist universe. Subsequently to sketching a modal analogue of four-dimensionalist supervenience in the next section, I shall, in the remaining four sections, embark on an extended critique of the temporal-parts account (and its temporal-counterparts analogue), showing that the account faces several real problems. The aim of this critique is to establish three-dimensionalist supervenience as superior to four-dimensionalist supervenience. In the case of the first two problems—the problem of spacetime asymmetry and the problem of ontological parsimony—the advantage of three-dimensionalism will be immediately

[7] Apparent spatial variation and spatial analogues of the ellipsis account and the subject-modifier account will be discussed in Sect. 4.3.

apparent. In the case of the third problem for the temporal-parts account (and the temporal-counterparts account)—the problem of predicational overkill—the advantage of three-dimensionalism will not become apparent until Chapter 5.

My critique of four-dimensionalist supervenience will be independent of certain standard lines of attack that feature in recent literature. Two classic arguments against four-dimensionalism are worth mentioning in this connection. First, there is the argument of the homogeneous discs (or spheres). Imagine two duplicate and homogeneous discs, one of which is stationary and the other rotating. These discs raise what can be put as a problem about supervenience: how can the difference between the discs be captured by the theory of temporal parts—that is, on which four-dimensionalist spacetime facts do the disc facts supervene? What this problem demands from the four-dimensionalist is an account of the genidentity relation, or unity relation, that holds together the temporal parts of an ordinary object. The objection to four-dimensionalism is that no satisfactory account of genidentity is available to explain the difference between the discs. Since the problem of the discs rules out analyses of genidentity as spatiotemporal or qualitative continuity, the now-standard view is that the analysis of genidentity should contain a causal component. Given the causal view of genidentity, one way of advancing the objection to four-dimensionalism is to claim that no account of the causal relation between an object's temporal parts can be given that is compatible with Humean Supervenience, the doctrine according to which all facts supervene on the distribution of local qualities across spacetime.[8]

Secondly, there is the argument from modal inductility. Intuitively, it seems right to say that a person could have existed longer than he or she actually did. The problem posed by this modal intuition, originally put forth by Peter van Inwagen (1990), is to capture the intuition in the light of the alleged consequence of four-dimensionalism that temporal parts, including the big temporal part that is the person itself, are modally inductile, meaning that they could not occupy a spacetime region that is different from the one they actually occupy. Van Inwagen's objection is that, in order to explain the modal intuition, four-dimensionalism must be combined with a particularly controversial account of modality *de re*—namely, counterpart theory (see Section 4.2 for a brief sketch of modal counterpart theory).

These two arguments have been much discussed, and various four-dimensionalist replies have been proposed. In the light of this discussion, the

[8] For discussion of the problem of the spinning disc, see Lewis (1986*b*: p. xiii, n. 5; 1999*b*), Robinson (1989), Zimmerman (1998*a*, 1999), Callendar (2001), Hawley (2001), and Sider (2001: 5.5).

objections from homogeneous rotating matter and from modal inductility seem answerable from a four-dimensionalist perspective.[9] Although there is still work to be done, I will not continue these debates, since I have no intention of defending four-dimensionalism; quite the opposite. Moreover, the main problem I focus on—the problem of predicational overkill—hits four-dimensionalism more directly than the aforementioned problems. The latter demand accounts of genidentity and *de re* modality. Such accounts are not part of the basic theory of temporal parts, which asserts only that ordinary objects have temporal parts unified by genidentity, no matter what exactly genidentity amounts to and no matter what modal properties objects and their temporal parts have. Unlike the disc argument and the modal argument, the argument to be given in Section 4.5 threatens the theory of temporal parts in its basic form.

4.2 TEMPORAL AND MODAL SUPERVENIENCE

Temporal predication has a modal analogue: Zoe was happy and Zoe might have been unhappy. There are widely recognized similarities between ordinary temporal discourse and ordinary modal discourse, discourse involving the idioms of possibility and necessity. The truths of ordinary temporal discourse logically supervene on truths of spacetime discourse. The question to be addressed in this section is whether there is an analogue of temporal supervenience with respect to ordinary modal truths. Since it is impossible to cover this topic in sufficient detail in the present context, I shall rest content with mere outlines.[10]

There is an obvious difficulty in speaking of supervenience in the modal case. Temporal facts supervene on non-temporal facts about spacetime. If modal facts supervene in a way that is analogous to temporal supervenience, then modal facts supervene on non-modal facts. But supervenience, as it was introduced in Chapter 1, is itself a modal concept. Therefore, this concept of supervenience cannot be employed in characterizing the relationship between the non-modal base facts and the ordinary modal facts. I am therefore using 'modal supervenience' in this section as a tag for the intuitive claim that non-modal base facts fully determine the ordinary modal facts, that, once the base facts are fixed, the modal facts follow.[11]

[9] Although not conclusive, the objections may still be viewed as favouring three-dimensionalism. See Sider (2001: 5.4, 5.5).

[10] Analogies between time and modality are also discussed in Dyke (1998) and Markosian (2001).

[11] In the often-assumed modal logic S5, any modal sentence—any sentence that is governed by a modal operator—is either necessarily true or necessarily false. Thus, a

My aim, to emphasize the point, is not to give a detailed account of certain issues in the metaphysics of modality, but only to point to an interesting analogy of these issues with certain issues in the metaphysics of time.

Ordinary modal facts are expressed by modal predications. Consider a modal predication of the form '*a* is possibly F'. What is the semantic status of the adverb 'possibly'? We are familiar with this sort of question from the discussion of the semantic status of tense (see Section 1.1). Tensers—those who 'take tense seriously'—say that grammatical tense in ordinary temporal predications is semantically irreducible, while detensers say that tense is semantically reducible. According to detenserism, '*a* was F' is equivalent to 'There is a past time at which *a* is F', where both occurrences of 'is' are tenseless. That is, tenses are explained in terms of times. In the modal case, a similar distinction may be drawn between those who 'take modality seriously' and claim that the modal idioms of possibility and necessity are semantically irreducible and those who claim that these idioms are semantically reducible. For the sake of analogy, we may call the first group *modalizers* and the second group *demodalizers*. The demodalizers hold that '*a* is possibly F' is equivalent to 'There is a possible world at which *a* is F'. That is, possibility and necessity are explained in terms of possible worlds. Let us set the modalizers aside and concentrate on the demodalizers.[12]

Temporal supervenience concerns the supervenience of facts relativized to times and expressed by sentences of the form '*a* exists at *t*' and '*a* is F at *t*' on spatiotemporal facts that are not relativized to times. Modal supervenience concerns the supervenience of facts relativized to possible worlds and expressed by sentences of the form '*a* exists at *w*' and '*a* is F at *w*' on spatiotemporal facts that are not relativized to worlds. Different spacetime ontologies are relevant for these two types of supervenience. Since temporal supervenience is supervenience on facts about *our* spacetime, a distinction that is relevant for temporal supervenience concerns what exists within our spacetime. One view is that a (classical) spacetime contains a manifold of hyperplanes of simultaneity. Another view is that a spacetime contains only a single such hyperplane. The first view is eternalism about spacetime, and the second view is presentism about spacetime (to be distinguished from eternalism and presentism about ordinary time; see Section 2.4). An ontological distinction that is relevant for modal supervenience concerns how many spacetimes exist, instead of how many hyperplanes exist within a spacetime. The first view on this issue is that there is only one spacetime,

modal fact—a true proposition expressed by a modal sentence—supervenes vacuously on any fact, if 'supervenience' is given the standard, modal interpretation. If 'supervenience' is given a non-modal interpretation, then modal facts can supervene non-vacuously.

[12] Modalism is put forth in Prior and Fine (1977) and Forbes (1985, 1989).

only a single universe, ours. The second view is that there are many spacetimes, a plurality of universes, in addition to ours. I will refer to the first of these views as *monism about spacetime* and to the second as *pluralism about spacetime.*[13] Monism about spacetime is the view held by almost every modal metaphysician except David Lewis and his followers.[14] According to Lewis, foreign spacetimes and their contents are of the same kind as our spacetime and its content. The difference between them is not a categorical one, but rather concerns the kinds of things that are located in different spacetimes.

Monists about spacetime deny that modal facts supervene on non-modal spatiotemporal reality. Most monists are ersatzists. Ersatzists say that, in addition to the one spacetime, there are countless abstract entities representing it. These are the possible worlds. The possible worlds, or ersatz worlds, represent the entire concrete world in complete detail; no possible world is silent on any matter of concrete fact. One possible world is actualized: it represents the concrete world correctly—that is, it represents the concrete world as it is. The rest are unactualized: they misrepresent the concrete world—that is, they represent it as it might have been. One form of ersatzism is linguistic ersatzism.[15] Linguistic ersatzism identifies possible worlds with maximal consistent sets of sentences. Correspondingly, possible worlds represent the concrete world by saying. For example, world *w* may represent Zoe as being happy by including a sentence that says that Zoe is happy. Ersatzist monism thus yields an account of modal predication along the following lines:

(T_3) 'At $w[F(a)]$' is true $\equiv$ w says that $F(a)$.

If modal supervenience holds, then spatiotemporal facts fully determine all modal facts. If monism is true, then there is only one spacetime system, one concrete world. Given ersatzism, this concrete world clearly does not fully determine all facts about possible worlds. How the concrete world is represented by ersatz worlds is not fully determined by how the concrete world is. How *a* really is does not fully determine how *w* says *a* is, just as how a politician is represented by the papers is not fully determined by how the politician really is. This is why ersatzist monism denies modal supervenience.

[13] Monism about spacetime should not be identified with actualism, as commonly understood, since some actualists leave it open whether many spacetimes are part of actuality.

[14] See Lewis (1986*a*).

[15] For more details on linguistic ersatzism and other forms of ersatzism, see Lewis (1986*a*: ch. 3).

There is an interesting temporal analogue of ersatzist monism: *ersatzist presentism.* The ersatzist presentist is a presentist about spacetime in holding that our spacetime is constituted by a single hyperplane of simultaneity. The ersatzist presentist is an ersatzist in holding that past and future times are not hyperplanes, as on the standard construal, but rather abstract representations of the instantaneous spacetime. This position presupposes a detenserist analysis of the tenses in terms of past and future times. A linguistic ersatz time is a maximal consistent set of temporally unmodified sentences of some language. The ersatz times represent the instantaneous spacetime in complete detail. One ersatz time is actualized: it represents the instantaneous spacetime correctly—that is, it represents the spacetime as it is. The rest are unactualized: they misrepresent the instantaneous spacetime—that is, they represent it as it was or will be. According to the ersatzist, the present is special in that it is the one ersatz time that is actualized. Given that presentism about ordinary time is the view that the present is special, the ersatzist is a presentist about ordinary time as well as about spacetime. Ersatz times represent the instantaneous spacetime by saying. For example, time *t* may represent Zoe as being happy by including a sentence that says that Zoe is happy. Ersatzist presentism thus yields an account of temporal predication along the following lines:

(T_4) 'At $t[F(a)]$' is true $\equiv$ t says that $F(a)$.

If temporal supervenience holds, then spatiotemporal facts fully determine all ordinary temporal facts. If presentism about spacetime is true, then spacetime is constituted by a single hyperplane. Given ersatzism, this instantaneous spacetime clearly does not fully determine all facts about ordinary time. How the instantaneous spacetime is represented by ersatz times is not fully determined by how the instantaneous spacetime is. How *a* really is does not fully determine how *t* says *a* is. This is why ersatzist presentism denies temporal supervenience.

The appearance of spacetime in the context of relativistic physics and the corresponding understanding of time and space as aspects of spacetime constitute a strong reason for expecting temporally unrelativized spatiotemporal facts to underlie ordinary temporal facts. Since there is no comparable reason for expecting non-modal to underlie modal facts, there is no imperative to explain modal supervenience, as there is in the case of temporal supervenience. In other words, there is no general problem of modal supervenience. Accordingly, the failure of ersatzist monism to sustain modal supervenience carries little weight, whereas the failure of ersatzist presentism to sustain temporal supervenience constitutes a major flaw of ersatzist presentism.

Pluralists about spacetime, unlike monists, may believe in modal supervenience and may explain it in a way that is analogous to the four-dimensionalists' explanation of temporal supervenience. With the purpose of giving a pluralist explanation of modal supervenience, modal predication may be given a representational account that differs from the ersatzists' account stated as (T_3), but that is the same as the representational account of temporal predication (T_2), with the *t*s replaced by *w*s. A sentence '*a* is F at *w*' is parsed as containing a sentence modifier 'At *w*' yielding 'At $w[F(a)]$'. This sentential operator is then interpreted as 'according to *w*', 'Ac *w*', and semantically reduced by the following definition:

(Def2′) Ac $w[F(a)] =_{df} w$ represents *a* as being F.

The right-hand side may be further analysed in terms of the now-familiar notion of a *representative*:

(Def3) *x* represents *a* as being F $=_{df}$ *a* has a representative in *x* that is F.

If 'At $w[F(a)]$' is interpreted as 'Ac $w[F(a)]$', and if (Def2′) and (Def3) are put together, then the right-hand side of (Def3) with '*x*' replaced by '*w*' gives the meaning of 'At $w[F(a)]$'. Since the right-hand side of (Def3) is a modally unmodified predication, modal predications may be given modally unmodified truth conditions:

(T_5) 'At $w[F(a)]$' is true $\equiv \exists x(\text{Rep}(x, a, w) \ \& \ F(x))$

where 'Rep(x, a, w)' is to be read as '*x* is *a*'s representative in *w*'.[16]

How is the representing done? That is, what is a representative of an object in a possible world? And what is a possible world in the pluralist framework? Let us take the last question first. We have already encountered the ersatzists who view possible worlds as abstract surrogates of concrete universes. The realists, following Lewis, view possible worlds as alternative concrete universes, different spacetimes. So the realists are pluralists.[17] Assuming the view that each possible world is a different spacetime, Lewis says that ordinary objects located in different possible worlds stand in counterpart relations to each other. The counterpart relation is a relation of comparative overall similarity: *a*'s counterparts located in a world *w* are

[16] The representational account of modal predication is an attempt to give precision to Lewis's idea of understanding modal discourse in terms of representation *de re*; see Lewis (1986*a*: 194–6).

[17] If all possible worlds are spacetime systems, and if the modal notions of possibility and necessity are understood in terms of possible worlds, then spacetime exists necessarily. Note further that pluralists may be ersatzists instead of realists. For actuality may be construed as a multiverse that contains many spacetimes, that is the only one of its kind, and that is represented in various ways by ersatz worlds.

all and only those things in *w* that are overall more similar to *a* than the other things.[18] Moreover, objects and their counterparts have properties simpliciter, meaning that instantiation is not relativized to a possible world (I am ignoring time here). Lewis's theory of modal counterparts is analogous to the theory of temporal counterparts sketched in Chapter 2, which in turn is a variant of the theory of temporal parts. The theory of temporal parts also has a modal analogue, according to which ordinary objects are trans-world individuals with modal parts, which Lewis contemplates and rejects.[19] Assuming that objects have counterparts in different possible worlds, we can say that:

(REP_{Real}) *x* is a representative of *a* in *w* iff *x* is a counterpart of *a* that is located in *w*.

This is a realist account of modal representation. With this account in place, modal supervenience can be explained. (REP_{Real}) functions as a bridge principle linking the representational account of modal predication with modal counterpart theory. The realist account of modal supervenience in a nutshell runs as follows:

(MS_{Real}) *a* is F at *w* iff *a* has a counterpart that is located in *w* and that is F simpliciter.

By this principle, the facts about the instantiation of properties by ordinary objects at possible worlds follow from facts about counterparts located in various spacetimes having various properties simpliciter. Since, according to the demodalizer, possible-world facts are equivalent to ordinary modal facts, such as that *a* is possibly F, the realist's construction of (MS_{Real}) from the representational account of modal predication and (REP_{Real}) explains how these ordinary modal facts supervene on non-modal reality.

So four-dimensionalist temporal supervenience has a modal analogue: realist modal supervenience. Just as the eternalist's manifold of three-dimensional slices comprising a spacetime fully determines all ordinary temporal facts, so the realist-pluralist's manifold of four-dimensional blocks comprising 'logical space' fully determines all ordinary modal facts. This analogy is not meant to speak in favour of four-dimensionalism. In fact, four-dimensionalist realists are more the exception than the rule—Lewis is one such exception. Nor is the analogy meant to speak against four-dimensionalism. Time and modality allow independent treatments. The main point of interest is rather that the idea of temporal supervenience and the techniques used by our four-dimensionalist to explain it have relevant applications beyond the temporal domain.

[18] See Lewis (1968, 1986*a*). [19] ibid. (1986*a*: 3.3).

4.3 SPATIAL SUPERVENIENCE AND SPACE–TIME ASYMMETRY

In addition to the problem of temporal supervenience, there is a problem of spatial supervenience. And since time and space are intimately linked, an account of temporal supervenience will have to be intimately linked with an account of spatial supervenience. It is, therefore, a further constraint on an account of temporal supervenience that it must be possible for such an account to be combined with an account of spatial supervenience. In what follows, I will first extend the four-dimensionalist temporal-parts account of temporal supervenience to cover spatial supervenience as well. Then I will make a first critical observation regarding the temporal-parts account. I shall argue that the four-dimensionalist disregards a fundamental asymmetry between ordinary space and time.

We have been guided by three questions in the metaphysics of time: What is time? How are ordinary objects in time? And how are ordinary properties instantiated in time? These questions have analogues in the metaphysics of space:

(*a*) What is space?
(*b*) How are ordinary objects in space?
(*c*) How are ordinary properties instantiated in space?

I will discuss each of these questions in relation to its temporal counterpart.

Ordinary space and spacetime

Question (*a*) concerns the nature of space. Space can be construed as ordinary space or as an aspect of spacetime, just as time can be construed as ordinary time or as an aspect of spacetime. Space and time as ordinarily conceived are distinct entities. Ordinary space is constituted by the places, regions, or areas—henceforth collectively to be referred to as 'places'—to which we are committed by our ordinary spatial discourse. Places are the intrinsically changeless and empirically inaccessible containers of everything concrete. Since they are empirically inaccessible, we pick them out via concrete things. Where is Suzie? In Paris. Paris is not a place in the strict sense currently under consideration, but rather a spread-out aggregate of ordinary objects, a city. Therefore, the city is not where Suzie is strictly speaking located if it is true to say that Suzie is located in Paris. The city is rather used to pick out the place where Suzie is meant to be located, so that the statement, fully spelled out, becomes 'Suzie is located somewhere in the place in which Paris is located'. Ordinary space, then, is a structure of three-dimensional places. Some features of ordinary space may be highlighted: places stand in spatial ordering relations to each other; for example, a place may be north of another place; places stand in metrical distance relations to each other; for example, a place may be three meters away from another place; and places have parts that are themselves places. So much for ordinary space.

As an alternative to the ordinary conception of space and time as distinct entities, space and time may be construed as different aspects of a single entity, spacetime. Space and time can be viewed as aspects of spacetime in the sense that certain spatial and temporal relations can be defined on the four-dimensional manifold of spacetime points and regions. Presupposing, for simplicity, a pre-relativistic and substantivalist account of spacetime, let us start with the familiar temporal equivalence relation, simultaneity, between spacetime points and regions that partitions the spacetime into hyperplanes of simultaneity. Each of these hyperplanes has the structure of Euclidean three-space, E^3, with its points and regions standing in various spatial ordering and spatial distance relations to each other. Furthermore, the points and regions of different hyperplanes may be partitioned into equivalence classes by a spatial equivalence relation that may be called *coincidence*.[20] Each of these spatial equivalence classes has the structure of the one-dimensional real line, E^1, with its points and regions standing in temporal ordering and temporal distance relations to each other. The intuitive significance of these equivalence classes of coincident points and regions will become apparent shortly. So much for a rough but sufficient sketch of the spatial as well as temporal aspect of spacetime.

Facts about the structure of ordinary time, about how objects are in ordinary time, and about how properties are instantiated in ordinary time logically supervene on facts about spacetime. This is the thesis of temporal supervenience, presenting the task of explaining how these three kinds of temporal fact supervene. Analogously, facts about the structure of ordinary space, about how objects are in ordinary space, and about how properties are instantiated in ordinary space logically supervene on facts about spacetime. This is the equally plausible thesis of spatial supervenience, presenting the task of explaining how these three kinds of spatial fact supervene. Since all statements about an ordinary object's relations to ordinary space are temporally modified, all ordinary spatial facts are temporal facts, and thus spatial supervenience is a particular form of temporal supervenience, a form

[20] This spatiotemporal notion of coincidence must be distinguished from the temporal notion of coincidence at a time. Distinct objects coincide at a time just in case they occupy the same place at the same time. This temporal notion of 'coincidence at a time' will become relevant in Sect. 5.6.

that we have not yet considered. So the problem of spatial supervenience is a part of the general problem of temporal supervenience. I will not address the supervenience of facts about the structure of ordinary space, and go straight to the question of how ordinary objects are in ordinary space.

Objects in space

From the perspective of ordinary space, the question how objects are in space is just as easily answered as the question how objects are in time is answered from the perspective of ordinary time. An object is in ordinary space by occupying a single place at a time, where this place is extended in three spatial dimensions. Since the place occupied by an object at a time has smaller parts, it is natural to say that the object has parts at that time that occupy these corresponding parts of the place. So the spatial parts of an object rest on the object's spatial extension. These are the familiar spatial parts of an ordinary object, the legs of a table, the arms of a person.

Occupation is here treated as a primitive relation. To occupy a place is to fit into the place perfectly, without leaving any gaps. Accordingly, a leg of a table does not occupy the place occupied by the table; nor does the table occupy the places occupied by its legs. Rather, the table and its legs each occupy a single place at a time. The spatiotemporal counterpart of occupying a place at a time, occupying a spacetime region simpliciter, was earlier employed in stating three-dimensionalism and four-dimensionalism, where the ordinary notion was appealed to as an intuitive guide to the technical notion. Below it will be shown how the two notions relate.

The next question is how objects are in space from the perspective of spacetime. In order to state different answers to this question, it will be helpful to pretend for a moment that spacetime is temporally unextended, that it is instantaneous, in the sense that there is only a single three-dimensional hyperplane of simultaneity. This restriction allows us to state spatiotemporal accounts of spatial extension independently of temporal extension. So how are objects located in spacetime thus narrowly conceived? There is, first:

> *Spatial three-dimensionalism*
> An ordinary object occupies a single spacetime region, and this region is extended in three spatial dimensions.

There are, further, *spatial two-*, *one-*, and *zero-dimensionalism*. Each of the latter three accounts says that an ordinary object occupies many spacetime regions, while disagreeing over the shape of these regions. The spatial two-dimensionalist says that the regions are extended in two spatial dimensions (planes); the one-dimensionalist says that they are extended in one dimension (lines); and the zero-dimensionalist says that they are spatially unextended (points). I shall take spatial three-dimensionalism to be the correct account. The other three views are merely of interest for comparison with the temporal case.

Spatial three-dimensionalism was stated for a temporally unextended spacetime. Let us now broaden the picture to capture a temporally extended spacetime. Recall that three-dimensionalism—the by now familiar temporal version—was glossed as the view that objects are extended along the three spatial dimensions but not along the temporal dimension, while four-dimensionalism was glossed as the view that objects are extended along all four dimensions. So three- and four-dimensionalism are views about spatial as well as temporal extension. These views may be explicitly factored into a temporal element and a spatial element as follows. We may first distinguish between *temporal one-dimensionalism* and *temporal zero-dimensionalism*:

> *Temporal one-dimensionalism*
> An ordinary object occupies a single spacetime region, and this region is extended in the temporal dimension.
>
> *Temporal zero-dimensionalism*
> An ordinary object occupies many spacetime regions, and these regions are temporally unextended, or instantaneous, and non-simultaneous.

For a region to be temporally extended is for it to have non-simultaneous parts, whereas for a region to be temporally unextended, or instantaneous, is for it to have only simultaneous parts. These two views remain silent on the spatial extension of the region or regions occupied. Temporal one-dimensionalism is analogous to spatial three-dimensionalism in that only a single, extended region is occupied. Temporal zero-dimensionalism is analogous to spatial two-, one-, and zero-dimensionalism in that many regions are occupied. Assuming that spatial three-dimensionalism is correct, this view may now be mixed with temporal one-dimensionalism and temporal zero-dimensionalism, respectively, to yield the following combinations:

> *Three-plus-one-dimensionalism*
> An ordinary object occupies a single spacetime region, and this region is extended in three spatial dimensions and the temporal dimension.
>
> *Three-plus-zero-dimensionalism*
> An ordinary object occupies many spacetime regions, and these regions are extended in three spatial dimensions but temporally unextended, or instantaneous, and non-simultaneous.

Three-plus-one-dimensionalism and three-plus-zero-dimensionalism are the familiar four-dimensionalism and three-dimensionalism with their spatial component made explicit. This spatial component is where the two views resemble each other, while differing with respect to their temporal component. Moreover, three-plus-one-dimensionalism asserts both spatial and temporal extension, and hence incorporates a space–time symmetry, whereas three-plus-zero-dimensionalism asserts spatial but not temporal extension, and hence incorporates a space–time asymmetry.

Three-plus-one dimensionalism was earlier developed to admit temporal parts. An instantaneous temporal part of an ordinary object *a* was characterized as a part of *a* that occupies a single instantaneous temporal part of the region occupied by *a*. An extended temporal part of *a* was characterized as a sum of non-simultaneous instantaneous temporal parts. The view that ordinary objects have temporal parts is based on temporal one-dimensionalism and the doctrine of arbitrary temporal parts, according to which, for every instantaneous temporal part of the region occupied by *a*, there is a part of *a* that occupies this region, and only this region. Three-plus-one dimensionalism may now be developed to include spatial parts as well. Let us say that an *instantaneous spatial part* of an object *a* is a *proper* part of an instantaneous temporal part of *a*—that is, a part of the temporal part that is distinct from the latter. A *temporally extended spatial part* of *a* is then a sum of non-simultaneous instantaneous spatial parts of *a*. Object *a*'s instantaneous spatial parts correspond to the familiar parts that *a* has at a time, such as its arms and legs. However, since we are presently talking at the level of spacetime, as opposed to ordinary space and time, it would be wrong to describe the instantaneous spatial parts of an object as the parts that the object has at a time. The three-plus-one dimensionalist now wants to say that, just as objects have arbitrary temporal parts, so they have arbitrary spatial parts. The simplest way of providing for both temporal and spatial parts is to generalize the doctrine of arbitrary temporal parts to yield the following doctrine of *arbitrary spatiotemporal parts*:

(ASTP) If an ordinary object x occupies a spacetime region R, then for every part R' of R there is a part x' of x such that (i) x' occupies R', and (ii) x' does not occupy any other spacetime region.

Combined with three-plus-one-dimensionalism, (ASTP) has the consequence that ordinary objects have both instantaneous and extended spatial and temporal parts. The result is an elegantly symmetrical picture of how objects are in spacetime: objects are extended and cut into parts both along the temporal dimension and along the spatial dimensions.

Explaining spatial supervenience

Having answered the question of how objects are in space both from the perspective of ordinary space and from the perspective of spacetime, the next question is how these answers are related. The general thesis of spatial supervenience is that all facts about ordinary space logically supervene on facts about spacetime. One particular thesis of spatial supervenience is the following: all facts concerning how an object is in ordinary space logically supervene on three-plus-one-dimensionalist facts about the spatiotemporal location of spatiotemporal parts of objects. I shall give a rough explanation of this thesis, focusing on the main facts about how an object is in ordinary space: the facts that an object occupies a place at a time, that it occupies a unique such place, and that the place is extended.[21]

Let us begin with a principle connecting places with spacetime regions: a place is an equivalence class of coincident spacetime regions, just as a time is an equivalence class of simultaneous spacetime regions. Moreover, a place is extended just in case its member-regions are extended. The intuitive significance of the equivalence relation of coincidence is thus that it 'bundles' spacetime regions into our familiar places. Now consider the sentence '*a* occupies *p* at *t*'. This seems to be a dyadic temporal predication, and so the three-plus-one-dimensionalist's natural move is to apply the representational account of temporal predication to it. Accordingly, the sentence has the logical form 'At t[O(a, p)]' and is interpreted as '*a* has a representative in *t* and *p* has a representative in *t* and *a*'s representative occupies *p*'s representative simpliciter'. We saw that a representative of *a* in *t* is a temporal part of *a* that occupies a region that is a member of *t*. This account obviously cannot be applied to the notion of being a representative of a place *p* in *t*, for *p* does not have temporal parts, and, even if it did have temporal parts, these temporal parts could not *occupy* members of *t*; they could only *be* members of *t*. The three-plus-one-dimensionalist is therefore advised to treat being a representative of *p* in *t* differently from being a representative of *a* in *t*, and hence to claim that there are different kinds of temporal representation. Being a representative of *p* in *t* may be construed as being a member of *p* as well as a member of *t*. The resulting

[21] Although an object's having a part at *t* is closely related to the object's occupying an extended place at *t*, facts about parthood at a time introduce nothing new from the perspective of supervenience. The sentence '*a* is a part of *b* at *t*' is a dyadic temporal predication that may be straightforwardly treated as an instance of the four-dimensionalist's supervenience principle (TS_{4D}): *a* is a part of *b* at *t* iff *a* has a temporal part, a_t, that occupies a member-region of *t* and *b* has a temporal part, b_t, that occupies a member-region of *t*, and a_t is a part of b_t simpliciter.

three-plus-one-dimensionalist account of occupying a place at a time looks like this:

(SO_{4D}) Necessarily, *a* occupies *p* at *t* iff *a* has a temporal part, a_t, that occupies a member-region of *t*, and *p* has a member-region, R_p, that is also a member of *t*, and a_t occupies R_p simpliciter.

This principle shows how the ordinary notion of occupying a place at a time is related to the technical notion of occupying a spacetime region simpliciter. What it says is that these notions can be treated as containing the same two-place predicate 'occupies', in one case temporally modified—At *t*[O(*a*, *p*)]—and in the other case temporally unmodified—SIMP[O(*a*, *R*)]. In this sense, occupation, as it was introduced at the level of spacetime, just is occupation as we know it. Given this principle, three-plus-one-dimensionalism together with the doctrine of arbitrary temporal parts—a restricted version of (ASTP)—entail that an ordinary object occupies a place at a time, that it occupies a unique such place, and that the place is extended. The core of spatial supervenience is thus explained.

This is the explanation of spatial supervenience by the three-plus-one dimensionalist. The three-plus-zero-dimensionalist offers a different supervenience thesis: all facts concerning how an object is in ordinary space logically supervene on three-plus-zero-dimensionalist facts about spatiotemporal location. How is this thesis to be explained? Consider again the sentence '*a* occupies *p* at *t*'. As we have just seen, the three-plus-one-dimensionalist treats such reports of spatial occupation along the lines of standard dyadic temporal predications. Since we have not yet found a workable three-plus-zero-dimensionalist account of temporal predication, we do not yet have the means to consider an analogous move in the three-plus-zero-dimensionalist case. As will become apparent in the next chapter, however, what I take to be the best three-plus-zero-dimensionalist account of temporal predication does not work for reports of spatial occupation anyway. This is not a drawback, though, because the three-plus-zero-dimensionalist has a simple alternative to (SO_{4D}). The idea is to deny that '*a* occupies *p* at *t*' is an ordinary temporal predication, and instead to treat it as an atemporal predication. This suggestion is on firm grounds, since '*a* occupies *p* at *t*' contains a spatial singular term, a designator of a place. Since places are abstract objects categorically distinct from ordinary material objects, there is reason for drawing a categorical distinction between '*a* is R to *p* at *t*' and '*a* is R to *b* at *t*', where *a* and *b* are material objects and *p* is a place. A report of spatial occupation may then be read as '*a* occupies *p*-at-*t*', or, formally, 'O(*a*, AT(*p*, *t*))', as opposed to 'At *t*[O(*a*, *p*)]'. Here 'AT(*p*, *t*)' is a singular term that works semantically as follows: 'AT' stands for a function that maps *p*, a set of spacetime regions, and *t*, also a set of spacetime regions, to the intersection of p and t. From the standard treatments of p and t, it is clear that the term 'AT(p, t)' designates a unique, instantaneous spacetime region. As in the case of (SO_{4D}), the predicate 'occupies', as it occurs in '*a* occupies a place p at a time' is the same predicate that occurs in '*a* occupies a spacetime region R simpliciter'. In fact, the present account of spatial occupation at a time provides the most direct connection with spatiotemporal occupation simpliciter. I take this to be a virtue of the account. Notice further that this treatment is not available to the three-plus-one-dimensionalist, since it straightforwardly entails that an ordinary object occupies many instantaneous spacetime regions—that is, it entails three-plus-zero-dimensionalism—on the basis that an ordinary object occupies various places at various times. Finally, it is easy to see that with this account of spatial occupation three-plus-zero-dimensionalism entails that an ordinary object occupies a place at a time, that it occupies a unique such place, and that the place is extended. This is how the three-plus-zero-dimensionalist can explain spatial supervenience.

Ordinary space–time asymmetry

We saw that an object is in ordinary space by occupying a single, extended place at a time. The temporal analogue of ordinary spatial extension, of occupying a three-dimensional place at a time, is occurring, or happening, at an interval, an 'extended' time. To persist through an interval—that is, to exist at each instant in the interval—is perhaps another way of understanding what it means to be extended in ordinary time, but not the understanding that is relevant here, the understanding that renders temporal extension analogous to spatial extension. If something occurs at an interval, then it takes time, begins at one instant and ends at another, and is in progress at instants in the interval. A party or a war can sensibly be said to take time, to begin at one instant and to end at another, and to be in progress at instants. But none of these things can be sensibly said of a table or a stone. The lives of tables and stones begin and end, not tables and stones themselves. So tables and stones do not occur at an interval. Tables and stones merely persist through an interval—they exist at each instant in the interval.[22]

We saw further that an object has parts at a time that correspond to parts of the extended place it occupies at that time. While objects clearly have

[22] It would be wrong to take the fact that I existed in 2004 as evidence for the claim that I existed at the interval that is 2004. As pointed out in Sect. 1.1, a natural analysis of temporal adverbials such as 'in 2004' is as predicates of instants, so that 'I existed in 2004' becomes '$\exists t$(I exist at t & $t \in$ 2004)'.

spatial parts in ordinary space, objects do not seem to have temporal parts in ordinary time, which corresponds to the impression that objects do not occur at extended times. Evidence for the claim that spatial parts of objects lack a temporal analogue in our ordinary thought and talk consists in the fact that, while we possess ordinary sortal terms, such as 'person' and 'table', for ordinary objects, and we possess ordinary sortal terms, such as 'arm' and 'leg', for spatial parts of objects, we possess no ordinary sortal terms for temporal parts of objects. It would be wrong to think that 'childhood' and 'adulthood' are sortal terms for temporal parts of a person, since these are sortal terms for temporal parts of the life of a person, where a life is an event that has a person, an object, as a subject, but is distinct from the person. So there is a fundamental asymmetry, according to our ordinary conception, concerning how objects are in space and time: objects occupy extended places, but do not occur at extended times; and objects have parts corresponding to parts of places, but do not have parts corresponding to 'parts' of times.

It is easy to see that this asymmetry causes trouble for the three-plus-one-dimensionalist. The latter asserts that in spacetime objects are temporally as well as spatially extended and have temporal as well as spatial parts. The spatiotemporal notion of occupying a temporally extended spacetime region obviously corresponds to the ordinary temporal notion of occurring at an interval, given that intervals are naturally correlated with temporally extended spacetime regions. Owing to this correspondence, we can expect the fact that an object occupies a temporally extended spacetime region to entail that the object occurs at a certain interval. It follows that, since the three-plus-one-dimensionalist claims that an ordinary object occupies a temporally extended spacetime region, the three-plus-one-dimensionalist is committed to the view that an ordinary object occurs at an interval. Moreover, the spatiotemporal definition of an 'instantaneous temporal part' given in Section 2.2 obviously corresponds to Sider's definition in ordinary temporal terms, first encountered in Section 2.3:

(TP) An object x is a *temporal part* of an object y at a time $t =_{df}$ (i) x exists at t, but only at t, (ii) x is a part of y at t, and (iii) x overlaps at t everything that is part of y at t.

Owing to this correspondence, we can expect the fact that an object has a temporal part, in the spatiotemporal sense, that occupies a member-region of t to entail that the object has a temporal part at t, in the sense of (TP). It follows that, since the three-plus-one-dimensionalist claims that an ordinary object has temporal parts in spacetime, the three-plus-one-dimensionalist is committed to the view that ordinary objects also have temporal parts in ordinary time. The result is that the three-plus-one-dimensionalist's space–time symmetry at the level of spacetime translates to the level of ordinary space and time, which clashes with our common conception of how objects are in space and time.

Remember that we are evaluating an account of how objects are in spacetime primarily by its success in capturing ordinary spatial and temporal facts. As we saw earlier, the three-plus-one-dimensionalist is well able to capture ordinary facts of persistence, change, and spatial occupation—ordinary facts that we want to admit. The problem is that the three-plus-one-dimensionalist 'creates' ordinary facts that we do not want to admit—namely, the facts that an ordinary object occurs at an interval, that it has temporal parts at instants, and hence that the way in which an object is in ordinary time is analogous to the way in which it is in ordinary space. No doubt, viewed only from the perspective of spacetime, three-plus-one-dimensionalism is pure elegance. But if ordinary space and time are taken seriously, as I think they should be, then the three-plus-one-dimensionalist design loses its elegance. This is also a good reason for formulating three-plus-one-dimensionalism and the theory of temporal parts in the first instance at the level of spacetime (as in Sections 2.1 and 2.2). For the spacetime framework is the theory's natural environment.[23]

It remains to be pointed out that three-plus-zero-dimensionalism is able to save the ordinary space–time asymmetry by giving ordinary spatial and temporal facts a spatiotemporal supervenience base that is itself characterized by an asymmetry between its spatial and temporal aspects. These considerations constitute the first stage in my extended argument for the conclusion that, if ordinary space and time are taken seriously, then three-plus-zero-dimensionalism is a better choice than three-plus-one-dimensionalism.

Spatial instantiation

The third entry on our list of questions in the metaphysics of space given at the beginning of this section is the question of how ordinary properties, such as shapes and colours, are instantiated in space. What is the answer to this question from the perspective of ordinary space? The first thing to say is that the objects that occupy places at times have properties at those times; likewise for the parts of these objects. Since these are just standard facts of temporal instantiation, they do not pose a new problem. However, temporal instantiation seems to have a spatial analogue that has not yet been considered: a road can be smooth in the valley and a pole can be

[23] Further instances of the problem that three-plus-one-dimensionalism implies unwanted ordinary facts will be discussed in Sect. 4.5.

green at the top. In short, just as *a* can be F at *t*, so *a* can be F at *p*. While spatial instantiation of this kind appears to yield ordinary facts whose supervenience on spatiotemporal facts needs to be explained, the truth is that spatial instantiation is a mirage. This will become clear by addressing a worry about spatial instantiation, a worry concerning, yet again, change.

The problem with spatial instantiation is usually stated as follows. Reports of change, such as 'Zoe is happy at t_1 and unhappy at t_2', have spatial analogues—for instance, 'the pole is green at the top and red at the bottom'. More generally, just as it is possible that *a* is F at t_1 and *a* is not F at t_2, so it is possible that *a* is F at p_1 and *a* is not F at p_2, where p_1 and p_2 are distinct places. Just as things can have a given property at one time and fail to have it at another, so things can have a given property at one place and fail to have it at another; temporal property-variation has a spatial analogue. Property-variation is change. It would be unnatural, however, to say that things change through space just as they change through time. So the problem is that there seems to be an analogy between space and time concerning change, which there should not be.[24]

Perhaps the problem rests on the false assumption that change is property-variation. Perhaps, that is, change is more than property-variation, to the effect that there is an extra bit that combines with temporal variation to yield change through time but fails to combine with spatial variation to yield change through space. Irrespective of what this sufficient condition might be, it is misplaced. We have a good intuitive grasp of what change amounts to. For instance, it seems obvious that something changes in colour over time just in case it has different colours at different times. Accordingly, it seems counter-intuitive to say that colour variation is not sufficient for change in colour, and that a further sophisticated condition is required to make it sufficient. This strategy threatens to deprive the concept of change of its status as an ordinary, pre-philosophical concept.

Furthermore, even if the concept of change were more complicated than we thought, the problem would remain untouched. This is so because the problem can be stated in terms of property-variation only, without even mentioning change. The standard formulation of the difficulty, given above, is little more than an expression of uneasiness about spatial variation. So what exactly is problematic about spatial variation? The problem arises from the plausible principle that variation implies persistence. In the temporal case, this principle is perfectly standard: temporal variation implies persistence through time; that *a* has different properties at different times implies that *a* exists at different times. Analogously, spatial variation in properties

[24] This unwanted space–time analogy is also discussed in Heller (1992). The conclusion to be reached below differs from Heller's.

implies spatial variation in location, or persistence through space; that *a* has different properties in different places at the same time implies that *a* occupies different places at the same time. The intuition that *a* occupies various places at the same time is not captured by *a*'s having various parts in various places at the same time. That *a* is different at different times implies that *a* is wholly at different times; and that *a* is different in different places implies that *a* is wholly in different places. This wording is not to be taken too seriously. It is just meant to emphasize that the natural reading of *a*'s existing at a time involves *a* itself and not just *a*'s parts, and the natural reading of *a*'s occupying a place likewise involves *a* itself and not just *a*'s parts. If a person travels back in time to meet her younger self, then it is perfectly sensible to say that, strictly speaking, the person itself has different heights in different places at the time at which she meets her younger self. However, if one end of a fence currently has one height and the other end of a fence currently has another height, then it is much less sensible to say that, strictly speaking, the fence itself currently has different heights in different places. The reason for this intuitive difference is that, while the person and her younger self occupy distinct places at the time of their meeting, the fence currently occupies a single place only. In other words, the person's variation in height across space corresponds with variation in location across space, whereas the fence's variation in height across space does not correspond with variation in location across space. So the problem about standard cases of spatial variation—cases involving no time travel—is that these cases involve no spatial persistence. Notice that the situation is unaffected by whether spatial variation counts as change or not.

This issue rests on the logico-semantic status of spatial predications with the surface form '*a* is F at *p* (at *t*)'. Prima facie, predications of this kind commit us to spatial instantiation, and hence to the troublesome spatial variation. In fact, however, there is no such commitment. We saw that treating the temporal modifier 'at *t*' in '*a* is F at *t*' as a sentence modifier stays close to the surface form of temporal predications, and accordingly yields genuine temporal variation. Similarly, if 'at *p*' in predications of the form '*a* is F at *p* (at *t*)' is treated as a sentence modifier, then we are stuck with spatial instantiation and variation. However, 'at *p*' may be treated by analogy with the ellipsis account or the subject-modifier account of temporal predication. I shall here focus on the ellipsis account because it is the simpler one, but I have no preference for the latter over the subject-modifier account. The ellipsis account of an apparent report of spatial variation is to read '*a* is F at p_1 (at *t*) and *a* is not F at p_2 (at *t*)' as short for '*a* [has a spatial part that] is F [and that is located] at p_1 (at *t*) and *a* [has a spatial part that] is not F [and that is located] at p_2 (at *t*)'. Recall that the temporal version of this account was earlier condemned for introducing temporal parts of ordinary objects at the level ordinary time (see Section 4.1). In the spatial case, the account is quite plausible, since ordinary space is full of spatial parts of ordinary objects. Thus, it is strictly and literally true that the leaf has the property of being green at t_1 and the property of being red at t_2, but it is not strictly and literally true that the pole has the property of being green at the top and the property of being red at the bottom. Strictly speaking, the top part of the pole has the property of being green and the bottom part of the pole has the property of being red. That is, strictly speaking, there is no spatial variation in this case. Our problem about spatial variation without spatial persistence is thus removed. Once an apparent report of spatial variation is fully spelled out, there is no spatial variation left, and hence it does not follow that an object is in different places at the same time. All that follows is that different parts of the object are in different places at the same time. Our ordinary spatial discourse, properly construed, therefore does not commit us to spatial instantiation. Correspondingly, spatial discourse does not raise a special problem of spatial-instantiation supervenience.

4.4 EXTREME FOUR-DIMENSIONALISM AND ONTOLOGICAL PARSIMONY

While the three-dimensionalist says that an ordinary object occupies many spacetime regions, the four-dimensionalist says that an ordinary object occupies a unique spacetime region. According to the four-dimensionalist, there is a one–one correspondence between ordinary objects and spacetime regions. Moreover, the mereological make-up of an ordinary object mirrors that of a temporally extended spacetime region. These correspondences seem to allow the four-dimensionalist to identify ordinary objects with temporally extended spacetime regions. No such identification is possible for the three-dimensionalist, since her occupation-relation is a one–many relation, whereas identity is a one–one relation. Before evaluating the prospects for the four-dimensionalist of identifying ordinary objects with extended spacetime regions, notice several structural consequences of this identification.

First, the simple thesis that an ordinary object *is* a temporally extended spacetime region replaces (4D). Given that extended spacetime regions have instantaneous and extended temporal parts, ordinary objects do as well. No separate theory of temporal parts of ordinary objects is required. Temporal parts of ordinary objects just fall out of the standard mereology of spacetime. Let us call this version of four-dimensionalism *extreme four-dimensionalism*, to be contrasted with *moderate four-dimensionalism*, which was stated as

(4D). Secondly, if moderate four-dimensionalism is replaced by extreme four-dimensionalism, then the moderate four-dimensionalist's account of spatiotemporal instantiation is naturally replaced by *regional instantiation*: instantaneous spacetime regions, temporal parts of ordinary objects, have ordinary properties simpliciter.

What speaks in favour of extreme four-dimensionalism is ontological parsimony. Start with moderate four-dimensionalism. It says that spacetime regions and their occupants are distinct. It also says that there is a one–one correspondence between ordinary objects and their parts, spatial and temporal alike, with spacetime regions and their parts. In other words, ordinary objects are the mirror image of temporally extended spacetime regions. Given this correspondence, the roles played by objects and their parts might as well be played by spacetime regions and their parts. These circumstances put the four-dimensionalist under pressure to identify ordinary objects with extended spacetime regions, for sustaining the distinction between objects and spacetime regions amounts to the admittance of a redundant category of things, and hence as unparsimonious ontology. As Sider, a friend of extreme four-dimensionalism, puts it, the identification 'is just crying out to be made'.[25] In short, methodological considerations concerning ontological economy commit the four-dimensionalist to the extreme version of her view. Three-dimensionalism ends up in a position between moderate and extreme four-dimensionalism. From the point of view of ontological economy, three-dimensionalism is better off than moderate four-dimensionalism, since there is no pressure on the three-dimensionalist to reduce ordinary objects to spacetime regions, given her one–many correspondence. The three-dimensionalist therefore has a clear reason within the spacetime conception to distinguish between ontological categories, a reason that the moderate four-dimensionalist lacks. But three-dimensionalism is worse off than extreme four-dimensionalism, since the latter is the simpler view: the extreme four-dimensionalist gets by with a single ontological category where the three-dimensionalist needs two. The four-dimensionalist ends up with the following deal: make extreme four-dimensionalism work and be ahead of three-dimensionalism; or fail to make it work, hence be committed to moderate four-dimensionalism, and be behind three-dimensionalism.

What speaks against extreme four-dimensionalism? It is obvious that extreme four-dimensionalism is a radical metaphysical view. Traditional substantivalist views about spacetime have in common the claim that the spacetime structure exists and has specific properties independently of the existence of any material objects. The relation between this substantival

[25] Sider (2001: 110).

spacetime and concrete matter is standardly conceived of as a relation of containment: the material world is 'contained in' spacetime. This containment-view of spacetime has the consequence that, on standard substantivalism, material objects are something over and above substantival spacetime; they are distinct from the spacetime regions they occupy. Extreme four-dimensionalism represents a radical departure from this tradition, in that it reduces the material world to spacetime itself: ordinary objects, such as persons and tables, *are* spacetime regions, and hence the material world is nothing over and above substantival spacetime.

That extreme four-dimensionalism, which carves the material world out of spacetime itself, is a radical departure from the substantivalist tradition is not objectionable in itself. However, this radical departure might be viewed as having objectionable consequences. What comes to mind immediately is extreme four-dimensionalism's apparent violation of common sense. Since persons are spacetime regions, spacetime regions are the way persons are and do the things persons do. For example, spacetime regions can be happy at a time. But surely spacetime regions are not the kinds of things that can meaningfully be said to be happy.

One reply to the objection that extreme four-dimensionalism clashes with common sense is to argue that a temporal predication about a spacetime region, such as the statement that a region is happy at a time, is not counter-intuitive. The claim that this statement is counter-intuitive, so one might argue, rests on intuitions about what spacetime regions are and what they are like. The status of such intuitions, however, is questionable. The notion of a spacetime region is a technical notion, a notion that plays certain theoretical roles, some of a physical, some of a metaphysical, nature. It is not an ordinary notion, a notion that plays a role in ordinary thought and talk. Thus, we cannot expect to have any direct intuitions in connection with this notion; we lack direct intuitive access to what spacetime regions are and to what they are like. And so the intuitive ground for denying that a spacetime region can be happy at a time is weak.

An alternative and stronger reply is to take the awkwardness of saying that a region can be happy at a time seriously and to concede that it would be false to ascribe happiness to a spacetime region at a time, while holding that property ascriptions are relativized to the way in which we think about an object. The idea is that, if a given object is thought of as a person, then it is meaningful to describe the object as being happy at a time, whereas, if the object is thought of as a spacetime region, then it is not meaningful to describe it as being happy at a time, despite the fact that the person is a spacetime region. The natural way of cashing out this proposal is by construing such predicates as 'is happy' as relativized to a sortal term, such as 'person' or 'spacetime region'. The details of sortally relativized predication will be discussed in Section 5.6, where we will encounter other cases that justify this apparatus.[26] These two replies, particularly the second one, go quite a way towards showing that extreme four-dimensionalism is not obviously counter-intuitive. So far, little stands in the way of extreme four-dimensionalism. But there is more to come.

To see why extreme four-dimensionalism is untenable, let us go back to spatial supervenience. Recall that the moderate four-dimensionalist explains the supervenience of ordinary facts of spatial occupation by means of the following principle:

(SO_{4D}) Necessarily, *a* occupies *p* at *t* iff *a* has a temporal part, a_t, that occupies a member-region of *t*, and *p* has a member-region, R_p, that is also a member of *t*, and a_t occupies R_p simpliciter.

This account was arrived at by applying the representational account of temporal predication to ordinary reports of spatial occupation, and by construing the notion of a temporal representative differently in the case of *a* and in the case of *p* (see Section 4.3). If spatial supervenience is to be explained in terms of extreme four-dimensionalism, then (SO_{4D}) must be modified, since *a* does not have temporal parts that occupy spacetime regions; *a*'s temporal parts *are* spacetime regions:

(SO_{e4D}) Necessarily, *a* occupies* *p* at *t* iff *a* has a temporal part, a_t, that is a member of *t*, and *p* has a member-region, R_p, that is also a member of *t*, and a_t occupies* R_p simpliciter.

According to (SO_{4D}), the predicate 'occupies' that occurs in ordinary reports of spatial occupation just is the standard spatiotemporal predicate. According to (SO_{e4D}), this correspondence is broken. On the right-hand side of (SO_{e4D}), a_t and R_p are one and the same spacetime region. Thus, a region bears the occupation*-relation to itself, which shows that occupation* is not the spatiotemporal occupation-relation that we are used to. According to extreme four-dimensionalism, there are simply no standard spatiotemporal facts of occupation, since there is nothing to occupy spacetime in the standard way. This feature of (SO_{e4D}) makes the latter less elegant than (SO_{4D}), but is not yet a reason for major concern.

[26] In the connection of defending extreme four-dimensionalism against the charge of being counter-intuitive, Sider makes the helpful comparison with the identity theory of mental states. If mental states are identical to physical states, then it follows that pains can be located in the brain, which seems counter-intuitive. The move usually made by identity theorists in response to this objection is analogous to the move made above: the ascription of properties to mental and physical states is relativized to ways of thinking about the states. See Sider (2001: 110–11).

The problem with (SO_{e4D}) appears when we consider the following straightforward generalization of (SO_{e4D}):

(TP_{e4D}) Necessarily, *a* is L to *p* at *t* iff *a* has a temporal part, a_t, that is a member of *t*, and *p* has a member-region, R_p, that is also a member of *t*, and a_t is L to R_p simpliciter.

This generalization is perfectly legitimate. Given that a place *p* can be the subject of a temporal predication, there must be a general account of temporal predications involving '*p*' and, correspondingly, a general explanation of how ordinary facts about *p* supervene on spacetime facts. Note further that (SO_{e4D}) and (TP_{e4D}) are arrived at via the representational account of temporal predication. We have seen reasons for believing that the latter is a good account for the four-dimensionalist to adopt. Even if the four-dimensionalist does not adopt the representational account as a way of getting to (TP_{e4D}), however, the need to explain spatial supervenience still commits her to (TP_{e4D}) or a close variant of the latter, assuming that '*a* is R to *p* at *t*' is treated as a temporal predication. Thus, if (TP_{e4D}) is problematic, then not only are extreme four-dimensionalists with representational inclinations affected, but all other extreme four-dimensionalists are as well.

The problem with (TP_{e4D}) is the following. Suppose that an object *a* has a temporal part, a_t, that is a member of a time *t*. Suppose further that a place *p* has a member-region, R_p, that is also a member of *t*. Suppose finally that a_t *is identical to* R_p simpliciter. As we know from the right-hand side of (SO_{e4D}), this scenario is possible. It follows by (TP_{e4D}), replacing the predicate 'is L to' by 'is identical to', that *a* is identical to *p* at *t*. But this result is unacceptable. An ordinary object occupies different places at different times. Therefore, to identify an object with its places at different times would be to adopt an absurdly strong form of temporary identity. The objection to (TP_{e4D}), then, is that it allows the identification of objects with their regions at the level of spacetime to translate to the level of ordinary space and time.[27]

[27] It would be too quick to add that, since an object is identical to a place at *t*, an object may occupy an object at *t*, and a place may occupy a place at *t*, which sounds false. I earlier mentioned the strategy of sortal relativization as a response to the objection that it is counter-intuitive to say that a spacetime region is happy at a time. Similarly, the extreme four-dimensionalist might say that the occupation-relation is sortally relativized, that it is sensible to how we conceive of its relata. Thus, the reply might continue, although an object is identical to its place at *t*, it does not follow that an object occupies an object at *t*, since a thing can be occupied only if we think of it as a place. It further does not follow that a place occupies a place at *t*, since a thing can occupy something only if we think of it as an object.

The only way to avoid this result is to deny that '*a* is L to *p* at *t*' is a temporal predication, and hence to drop (SO_{e4D}) and (TP_{e4D}). How this might be done was shown in Section 4.3, where this strategy was introduced as the heart of the three-dimensionalist account of spatial supervenience. The basic idea is to formalize '*a* is L to *p* at *t*' as 'L(*a*, AT(*p*, *t*))'. Here the term 'AT(*p*, *t*)' designates an instantaneous spacetime region—namely, the intersection of *p* and *t*.[28] Since *a*, an extended spacetime region, is not identical with any instantaneous region, it does not follow that *a* is, at *t*, identical to the place *a* occupies at *t*, which was the problem with (TP_{e4D}).

A new problem arises, however. Consider first the extreme four-dimensionalist's variant of the dyadic version of (TS_{4D}): *a* is R to *b* at *t* iff *a* has a temporal part, a_t, that is a member of *t*, and *b* has a temporal part, b_t, that is a member of *t*, and a_t is R to b_t simpliciter. Note that the extreme four-dimensionalist denies the application of this principle to dyadic temporal predications about an object and a place, such as '*a* occupies *p* at *t*', but accepts the application of the principle to dyadic temporal predications about two objects with temporal parts. The problem can now be stated. According to extreme four-dimensionalism, both an instantaneous temporal part of an object and the referent of the complex singular term 'AT(*p*, *t*)' are instantaneous spacetime regions. Thus, the temporal part of *a* that exists at *t* is identical to AT(*p*, *t*), where *p* is the place *a* occupies at *t*. It is possible, then, that *a* has a temporal part, a_t, that is a member of *t*, and AT(*p*, *t*) has a temporal part, namely itself, that is a member of *t*, and a_t is identical to AT(*p*, *t*). By the extreme four-dimensionalist's dyadic version of (TS_{4D}) stated above, it follows that *a* is identical to AT(*p*, *t*) at *t*, which means in ordinary language that object *a* is identical to place *p* at *t*, at *t*. This result is problematic for two reasons. First, we end up with a questionable identification of ordinary objects with the places they occupy at different times. As in the case of (TP_{e4D}), the identification of objects with their regions at the level of spacetime is allowed to translate to the level of ordinary space and time. Secondly and more importantly, we end up with a true predication containing a non-vacuous iteration of the temporal modifier 'at *t*'. As pointed out in Section 3.3, we may think of the time specified by a temporal modifier as an answer to the question 'When does the fact that *a* is F, or the fact *a* is *R* to *b*, obtain?' Then it is obvious that, once a time is specified, there is no further question to be answered by another temporal operator; it is not sensible to ask, 'When does the fact that *a* is F

[28] Application of this reading to '*a* occupies *p* at *t*' yields: O(*a*, AT(*p*, *t*)). Since *a* occupies different places at different times, the occupation-relation between *a* and different spacetime regions such as AT(*p*, *t*) had better not be the standard relation, or else three-dimensionalism follows.

at t, or the fact that a is R to b at t, obtain?' As a consequence, a temporal modifier that applies to a true sentence that is already temporally modified must be vacuous. Since on the current extreme-four-dimensionalist account of temporal predication the sentence 'a is identical to p at t, at t'' can be true while neither operator is vacuous, this account of temporal predication cannot be correct. I conclude that the proposal to read 'a is L to p at t' as 'L(a, AT(p, t))' fares no better than (TP_{e4D}).

As a last point, one might wonder whether the extreme four-dimensionalist could not simply deny that 'p' and 't', in 'a is L to p at t', designate a place and a time, respectively. Perhaps, that is, such statements require a more substantial reformulation than has been considered above. The reason why the extreme four-dimensionalist is committed to places and times is that she is, by definition, a substantivalist about spacetime. Substantivalism is the view that there are spacetime points and regions, which has the natural correlate at the level of ordinary space and time that there are places and times, standardly construed as certain sets of spacetime points and regions. Thus, to deny that there are places and times is a form of relationism about spacetime, which I decided at the beginning to leave out of the picture entirely.

What the argument given above shows is that each of the extreme four-dimensionalist's explanations of spatial supervenience—that is, explanations of how ordinary spatial facts, facts about places, supervene on spatiotemporal facts—has unacceptable consequences. This result is of relevance for the dispute between four-dimensionalists and three-dimensionalists. Earlier considerations of ontological economy led to the observation that three-dimensionalism occupies a position between moderate and extreme four-dimensionalism: three-dimensionalism is better off than moderate four-dimensionalism, since there is no pressure on the three-dimensionalist to reduce ordinary objects to spacetime regions; but three-dimensionalism is worse off than extreme four-dimensionalism, since the latter is the simpler view. We know now that the ontological desert of extreme four-dimensionalism is uninhabitable. The four-dimensionalist is therefore committed to the moderate version of her view, which means that she is committed to a redundant ontological category. The discussion of extreme four-dimensionalism thus provides a further reason in favour of three-dimensionalism.

4.5 PREDICATIONAL OVERKILL

The four-dimensionalist claims that ordinary facts of temporal persistence and temporal instantiation logically supervene on spatiotemporal facts concerning the properties of temporal parts of ordinary objects. The explanatory link that the four-dimensionalist needs to bridge these different kinds of fact is provided by the following principle:

(TS_{4D}) Necessarily, a is F at t iff a has a temporal part that is F simpliciter and that occupies a spacetime region that is a member of t.

In this section, I will criticize (TS_{4D}). Before I start the critique, a word is in order on what such a criticism shows, if it is correct. Recall that (TS_{4D}) was derived in several steps, including the development of the representational account of temporal predication and of the temporal-parts account of temporal representation. Together these accounts pave the way to (TS_{4D}), and hence explain four-dimensionalist supervenience. (TS_{4D}) itself, however, does not presuppose the representational account of temporal predication and the temporal-parts account of temporal representation. While the representational explanation is very promising, there might be another plausible way of deriving (TS_{4D}). (TS_{4D}) itself, on the other hand, is compulsory. It is the link that the four-dimensionalist needs to back up her supervenience theses, no matter how this link is explained. Therefore, to criticize (TS_{4D}) is not just to criticize some account of temporal predication. Since (TS_{4D}) is neutral on which account of temporal predication is active in the background, to criticize (TS_{4D}) is to raise doubts about the very possibility for the four-dimensionalist to explain temporal supervenience. And, since temporal supervenience must be explained, to criticize (TS_{4D}) is to criticize four-dimensionalism and the theory of temporal parts themselves.

The temporal-parts account of temporal supervenience says that a is F at t iff a has a temporal part that occupies a member-region of t and that is F. An important consequence of this account is that temporal predication is *closed under the temporal-part relation* in the following sense. Since all that is required to be F at t is to have a temporal part, a_t, that occupies a member-region of t and that is F, every object that has a_t as a temporal part is F at t. Hence, if a is F at t, then every object that overlaps with a in that it has a_t as a temporal part is also F at t. Closure under parthood has various counter-intuitive consequences. I call the problems raised by closure problems of *predicational overkill.*

The first difficulty with closure under parthood concerns uniqueness. Consider the following example:

(1) Zoe, and only Zoe, is happy at t.

This must surely be possible. By the temporal-parts account, Zoe has a temporal part, z_t, that occupies a member-region of t and that is happy (simpliciter). Given that Zoe is a person and has infinitely many temporal parts, she has infinitely many proper temporal parts that also have z_t as a temporal part. Focus on one of them and call it 'Zoe-minus'. Then Zoe-minus has a temporal part, z_t, that occupies a member-region of t and that is happy. By closure under parthood, it follows that Zoe-minus is happy at t, which contradicts the assumption that Zoe is the only one who is happy at t because Zoe-minus is a proper part of Zoe and therefore distinct from Zoe. Hence, (1) is impossible. Or suppose, to give another example, that Zoe is a wife of Billy at t. By the dyadic version of (TS_{4D}), Zoe has a temporal part, z_t, at t and Billy has a temporal part, b_t, at t and z_t is a wife of b_t.[29] Given that Billy is a person, he has infinitely many proper temporal parts that also have b_t as a temporal part. By parthood closure, it follows that Zoe is a wife not only of Billy at t but of infinitely many of his temporal parts as well. The theory of temporal parts thus entails polygamy on a massive scale.

What this seems to show is that closure under parthood renders it impossible that an ordinary, persisting object could be the *only* object that has a certain property at a time or that bears a certain relation to another object at a time. There will necessarily be infinitely many temporal parts of the object that also have that property or relation at that time. If the temporal-parts account of temporal supervenience is correct, then too many things have properties and relations at times. This is predicational overkill.[30]

What can the friend of temporal parts do to remove the problem with (1)? A good start is to point out that there are two different readings of (1). Informally, (1) can be read as 'Zoe, and only Zoe, is such that at t she is happy' or as 'At t, Zoe, and only Zoe, is happy'. Formally:

(2) At t[Happy(z)] & $\forall x$(At t[Happy(x)] $\supset x = z$)
(3) At t[Happy(z) & $\forall x$(Happy(x) $\supset x = z$)]

The difference between (2) and (3) concerns the position(s) of the temporal modifier 'At t'. In (2), the universal quantifier and '$x = z$' lie outside the

[29] The dyadic version of (TS_{4D}) says that, necessarily, a is R to b at t iff a has a temporal part, a_t, that occupies a member-region of t & b has a temporal part, b_t, that occupies a member-region of t & a_t is R to b_t.

[30] There is a superficial resemblance between this problem and Peter Unger's 'problem of the many'. In both cases there are many things where we thought there was just one. But the two problems are independent in the sense that, if one is resolved, the other remains. Lewis's 'almost-one' solution, for instance, cannot be applied to the present problem. For we cannot say, in Zoe's case, that the many objects sharing her temporal part at t are 'almost one', since only one of them, Zoe, is an ordinary object, a person. We can say, however, that the many objects sharing her temporal part at t are identical at t (see the first reply in the text). But this is still different from the 'almost-one' solution, since now the many objects are completely one at t, not almost one. Conversely, the identity-at-a-time reply to the present problem cannot be applied to the problem of the many. See Unger (1980) and Lewis (1993).

scope of 'At t', and so (2) implies '$\exists!x$(At t[Happy(x)])'. In (3), the universal quantifier and '$x = z$' lie inside the scope of 'At t', and so (3) implies 'At t[$\exists!x$(Happy(x))]'.

The reading of (1) that leads to trouble is (2). The problem arises because, if Zoe is happy at t, then Zoe has proper temporal parts that are also happy at t and yet distinct simpliciter from Zoe. The friend of temporal parts may thus claim that (2) should be rejected in favour of (3). The reason why no trouble arises from (3) is that, although an object x that is happy at t is distinct simpliciter from a proper temporal part y that is also happy at t, x is identical to y at t. For instance, Zoe and Zoe-minus share a temporal part, z_t, that occupies a member-region of t and that is identical to itself. By the dyadic version of (TS_{4D}), it follows that Zoe is identical to Zoe-minus at t. The point is that, if we count by temporally unrelativized identity, identity simpliciter, then infinitely many things are happy at t if Zoe is happy at t. But if we count by the temporally relativized identity, identity at a time, then Zoe is the only one who is happy at t, because infinitely many things that are distinct simpliciter can be identical at t.[31]

This response seems appropriate in cases involving a single temporal modifier. But now consider a similar case involving several modifiers:

(4) Zoe, and only Zoe, is happy at t_1 and sad at t_2.

This is an example of what I earlier called a *cross-temporal predication* (see Section 3.3). In such a predication we talk about an object as it is over a certain period of time, in which we trace an object through time. Cross-temporal predications are of central importance for the present discussion. While (1) is an example of counting at a time, (4) is an example of counting across time. (4) is not saying that only Zoe is happy at t_1 and only Zoe is sad at t_2. (4) is rather saying that Zoe is the only one who is both happy at t_1 and sad at t_2. (4) must therefore be read as 'Zoe, and only Zoe, is such that at t_1 (she is happy) and at t_2 (she is sad)'. Or formally:

(5) At t_1 [Happy(z)] & At t_2 [Sad(z)] & $\forall x$(At t_1 [Happy(x)] & At t_2 [Sad(x)] $\supset x = z$)

As in the case of (2), the universal quantifier and the identity '$x = z$' lie outside the scope of the temporal operator 'At t', and so (5) implies '$\exists!x$(At t_1 [Happy(x)] & At t_2 [Sad(x)])'. Accordingly, the temporal-parts account renders (5) false just as it renders (2) false. Since temporal predication is

[31] In a related context, Lewis distinguishes between the simple relation of identity and the weaker relation of identity-at-t: Persons are identical-at-t iff they have stages (temporal parts) at t that are identical. Then Lewis suggests that we sometimes count by identity and sometimes by identity-at-t. See Lewis (1976: 63–4).

closed under parthood, if Zoe is happy at t_1 and sad at t_2, then Zoe has proper temporal parts that are also happy at t_1 and sad at t_2 and yet distinct simpliciter from her. Thus, it seems that the temporal-parts account makes (4) impossible.

As an attempt to overcome the difficulty with (4), the friend of temporal parts might suggest that (4) contains an implicit temporal modifier 'At t_0' yielding:

(6) At t_0[Zoe, and only Zoe, is happy at t_1 and sad at t_2]

The sentence modified by 'At t_0' may be unpacked as (5), which has the consequence that '$x = z$' lies inside the scope of 'At t_0'. Analogously to the case of (3), (6) seems to come out true on the temporal-parts account because all the things that are happy at t_1 and sad at t_2 but distinct simpliciter from Zoe are identical to Zoe at t_0.

Apart from being completely ad hoc, this suggestion is unacceptable because temporal predications such as 'At t_1[Happy(z)]' end up inside the scope of a further temporal operator 'At t_0', and hence as multiply temporally modified. As pointed out before, a temporal modifier 'At t' that applies to a sentence that is already temporally modified has no effect; the operator is vacuous. If we think of the time specified by a temporal modifier as an answer to the question 'When does the fact that *a* is F obtain?', then it is obvious that, once a time is specified, there is no further question to be answered by any other evaluation time; it is not sensible to ask 'When does the fact that *a* is F at *t* obtain?' Since the outermost temporal operator in (6), 'At t_0', is vacuous, this operator may be dropped from (6), and hence it does nothing to remove the original difficulty with (4). A more technical reason for holding that 'At t_0' in (6) is vacuous is that (TS_{4D}) renders every multiply temporally modified statement false. Consider the simple '((*a* is F) at t_1) at t_2'. The statement governed by 'at t_2', '(*a* is F) at t_1', is true iff *a* has a temporal part that occupies a member-region of t_1 that is F. Accordingly, the whole statement is true iff *a*'s temporal part at t_1 has a temporal part at t_2 that is F. Since an instantaneous temporal part that occupies a member-region of one time cannot have an instantaneous temporal part that occupies a member-region of another time, the statement cannot be true.

Another attempt of dealing with cross-temporal counting might be to read (4) as saying that Zoe is happy at t_1 and at t_2, and everything that is happy at t_1 and at t_2 is identical to Zoe at t_1 and at t_2. As on the previous proposal, what makes this reading of (4) compatible with the temporal-parts account is that the predication of identity is temporally modified. What makes the current proposal better than the previous one is that iteration of temporal modifiers is avoided; the strategy is to modify not the whole sentence but only the predication of identity.

In order to see where this proposal goes wrong, consider first the sentence 'At least two things are F between t_1 and t_2'. This sentence says that there is an x that is F between t_1 and t_2, and there is a y that is F between t_1 and t_2, and x is distinct from y simpliciter. The predication of distinctness in this sentence must be temporally unmodified, because we cannot assume that there is any time at which x and y both exist. Perhaps x exists throughout one part of the interval, whereas y exists throughout another, wholly distinct part of the interval. If x and y were distinct at some time t, however, x and y would both exist at t. Suppose now that Zoe is F between t_1 and t_2. By the theory of temporal parts and closure under parthood, Zoe has a temporal part that is also F between t_1 and t_2 and distinct simpliciter from Zoe. Thus, at least two things are F between t_1 and t_2 if Zoe is. Next, consider the sentence 'Exactly one thing is F between t_1 and t_2'. Following the current account of cross-temporal counting, this sentence is to be understood as saying that there is an x that is F between t_1 and t_2, and for all y, if y is F between t_1 and t_2, then y is identical to x at all times at which x exists between t_1 and t_2. The point of the proposal is to make it possible for four-dimensional objects such as Zoe to fulfil the conditions for being the only thing that is F between t_1 and t_2. Supposing that Zoe does fulfil these conditions, it is still the case that at least two things are F between t_1 and t_2. For, even if every proper temporal part of Zoe that is F between t_1 and t_2 is identical to Zoe between t_1 and t_2, each of these proper temporal parts is still distinct simpliciter from Zoe. The contradictory result is that at least two things are F between t_1 and t_2, and exactly one thing is F between t_1 and t_2. The construal of cross-temporal counting under consideration is therefore incorrect. In order to avoid the problematic result, the sentence 'Exactly one thing is F between t_1 and t_2' must be construed as predicating identity simpliciter: there is an x that is F between t_1 and t_2, and for all y, if y is F between t_1 and t_2, then y is identical to x simpliciter. This brings us back to reading (5) of (4), which was shown to be incompatible with the temporal-parts account.

A further difficulty with closure under parthood concerns cross-temporal continuity. Consider the following example:

(7) Everything that is happy at t_1 is still happy at t_2.

Intuitively, there is no doubt that this statement can be (non-vacuously) true. But take an arbitrary object x such that x is happy at t_1. Since temporal predication is closed under parthood, x_{t_1} is also happy at t_1, x_{t_1} being x's instantaneous temporal part occupying a member-region of t_1. For an object to exist at t_2, it needs to have a temporal part that occupies a member-region of t_2. The instantaneous x_{t_1} fails to have such a temporal part and so does not exist at t_2. Accordingly, x_{t_1} is not happy at t_2. But then it follows that, for every object that is happy at t_1 and at t_2, there is an object that is also happy at t_1 but fails to be happy at t_2. Hence, (7) cannot be (non-vacuously) true. The result is that, just as closure under parthood renders it impossible that a persisting object be the only object that has a certain property at one time and a different property at another time, so closure renders it impossible that every object that has a certain property at one time still has that property at another time. Both results are unacceptable.

Closure under parthood leads to predicational overkill; four-dimensionalist supervenience allows too many things to have ordinary properties at times. In order to avoid predicational overkill, temporal predication must be restricted. But how? A restriction must not be arbitrary; a restriction must have a principled basis. I can think of one idea that promises to satisfy these criteria and to help the four-dimensionalist's cause. The idea is that ordinary temporal discourse is *sortally restricted*; ordinary temporal discourse, or at least a part of it, is true only of ordinary objects, of objects that fall under some ordinary sort, such as being a person. Sortal restriction should give the four-dimensionalist hope to avoid the trouble with closure because she could claim that, of all the things that, by closure, come out as being happy at t if Zoe is happy at t, only one is a person, Zoe. In what follows, I will spell out and evaluate this strategy in detail and show that it cannot do the job required by the four-dimensionalist. The starting point in the discussion is to clarify different roles that sorts can play in the theory of temporal parts.

Sorting the theory of temporal parts

The third thesis of the theory of temporal parts, (T3), says that an ordinary object is a sum of temporal parts that is maximal under some unity relation (see Section 2.2). A clearer way of putting this is to say that our ordinary singular terms designate sums of temporal parts that are maximal under some unity relation. This thesis leaves open what falls under an ordinary sort, such as being a person or being a chair. More precisely, the thesis leaves open how sorts are distributed across spacetime, assuming the theory of temporal parts (which is stated at the level of spacetime). There are two relevant options: (i) maximal unified sums of instantaneous temporal parts fall under ordinary sorts simpliciter; or (ii) instantaneous temporal parts of maximal unified sums fall under ordinary sorts simpliciter.[32] These two options have very different consequences. According to (TS_{4D}), a property

[32] The thesis that ordinary singular terms designate spacetime worms and that instantaneous stages of these worms fall under ordinary sorts simpliciter differs from the theory of temporal counterparts, according to which ordinary singular terms designate stages (see Sect. 4.6).

that is had by an object at a time is had by an instantaneous temporal part of that object simpliciter. Option (ii) construes instantaneous temporal parts as persons, and hence allows that sums of such temporal parts, the referents of our names, are persons at times. In short, (ii) admits *temporal sorts*. Option (i), on the other hand, does not allow instantaneous temporal parts to be persons, and hence does not, by (TS_{4D}), allow that maximal sums of temporal parts are persons at times. In short, (i) does not countenance temporal sorts, but only *atemporal sorts*.

The four-dimensionalist is committed to option (ii) because option (i) is misguided. As I have stressed repeatedly, there is a fundamental difference between describing the world in the conceptual framework of ordinary time and describing the world in the conceptual framework of spacetime. In the framework of ordinary time, objects are described from different temporal perspectives, in a series of temporal snapshots. That is, in ordinary time an object is described as having its properties at different times, corresponding to the fact that an object exists at different times. This is our ordinary, temporally relativized perspective on the world of objects.[33] We obviously group objects into sorts from this ordinary perspective. We sort objects into persons or chairs and describe them as being happy or wooden. All this is done in the common framework of ordinary time, and hence objects can be persons or chairs at a time, just as they can be happy or wooden at a time. This is not to say that an object can change with respect to being a person or a chair in the same way in which it can change with respect to being happy or wooden. The point is merely that our ordinary attributions of sortal terms to material objects are temporally relativized. As a consequence, temporal sorts as postulated in (ii) must be admitted.

If temporal sorts as postulated in (ii) must be admitted, then atemporal sorts as postulated in (i) must be rejected. In the framework of spacetime, objects are described from an atemporal perspective. They are described *sub specie aeternitatis*: nothing has its properties at a time; everything has its properties simpliciter. The spacetime picture and the ordinary picture are disjoint. When we view the world from the perspective of ordinary time, which is our common angle, then the spacetime picture is hidden from us. Thus, if, as option (i) says, objects can be persons or chairs only simpliciter, and hence objects cannot be persons or chairs at a time, then persons or chairs are hidden from our ordinary view; then we cannot describe anything as a person or a chair in ordinary, temporally sensitive language.

[33] The description of objects in terms of different instantaneous snapshots corresponds to the fact that objects exist at instants. Ordinary events, such as parties, on the other hand, occur at temporally extended intervals. Accordingly, events are not described in a series of instantaneous snapshots.

Moreover, to say that we do ordinarily describe an object as being a person or chair simpliciter is to confuse the technical language of spacetime with the everyday language of ordinary time.

This point may be illustrated by means of an analogy. Consider a world populated by objects that are extended in three spatial dimensions, and take one object in that world, a sphere. Now imagine a world with only two spatial dimensions. In this world—let us call it *Flatland*—objects have only two spatial dimensions and everything is confined to a surface. Assume further that the original three-dimensional world and the two-dimensional Flatland are intimately related in the following way: Flatland constitutes a certain two-dimensional perspective on the three-dimensional world, which may be represented by a plane intersecting a sphere, as in Figure 4.

While there is a sphere from the three-dimensional perspective, there is only a disc—a slice of the sphere—from the perspective of Flatland. Let us assume that the inhabitants of three-dimensional space and the Flatlanders speak about exactly the same objects—their names designate the same things—so that they describe the same things in different ways. This is a faithful analogy of the relationship between spacetime and ordinary time, according to four-dimensionalism. The three-dimensional perspective is analogous to the spacetime perspective and the Flatlanders' perspective is analogous to ordinary time, where Flatland corresponds to a single time. So imagine ordinary time as a series of ordered Flatlands. Consider, then, the role of sorts in the Flatland/3Dland scenario. One and the same object is described as a sphere from the three-dimensional perspective and as a disc from the two-dimensional perspective. The important point for present purposes is that the only sorts that the Flatlanders have access to are the sorts under which slices of the three-dimensional object fall.

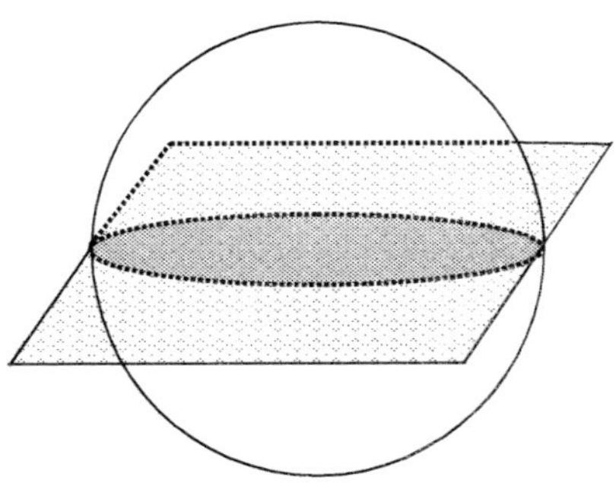

3Dland vs. Flatland

Figure 4

Hence, only disc-talk is admitted in Flatland. The sort under which the three-dimensional object falls itself is inaccessible to the Flatlanders. The analogous point in the ordinary time/spacetime case is that the only sorts that are accessible from ordinary time are the sorts under which slices of a four-dimensional object fall. The sort under which the four-dimensional object falls itself is inaccessible to our ordinary view. The four-dimensionalist should therefore say that ordinary sorts apply to instantaneous temporal parts of four-dimensional objects, and hence turn up as temporal sorts in ordinary time, whereas four-dimensional objects only fall under non-ordinary sorts, such as being a spacetime worm. The temporal-sorts view is the view the four-dimensionalist should adopt.

To say that our ordinary attributions of sortal terms are temporally relativized is not to claim that all our ordinary talk involving sortal terms clearly supports the temporal-sorts view. What is true, though, is that some of our ordinary talk involving sortal terms clearly supports the temporal-sorts view, while talk that is neutral may be read in a way that is consistent with the temporal-sorts view. In such sentences as 'Socrates was a person' the commitment to temporal sorts is obvious. For more subtle ways of invoking temporal sorts, consider the following pair of sentences:

(8) At each time between t_1 and t_2, there are two persons who are happy.
(9) There are four persons who are happy between t_1 and t_2.

There is a natural reading of (8) as counting persons at each time between t_1 and t_2—that is, as counting persons inside the scope of the temporal operators. And there is a natural reading of (9) as counting persons across the interval from t_1 to t_2—that is, as counting outside the scope of the temporal operators. On these readings, (8) and (9) can both be true. The reading of (8) as counting persons at each time between t_1 and t_2 puts the numerical quantifier inside the scope of 'at each time between t_1 and t_2'. Since the sortal 'person' lies inside the scope of the numerical quantifier—we are counting persons—it follows that (8), on the natural reading under consideration, requires the ascription of the sortal term 'person' to be temporally modified. Thus, (8) constitutes clear evidence for the temporal-sorts view. The reading of (9) as counting persons across the interval from t_1 to t_2 puts the numerical quantifier outside the scope of 'between t_1 and t_2'. This reading is neutral about the position of the sortal 'person'. One option is to place the sortal outside the scope of 'between t_1 and t_2', as in (9′); the other option is to place the sortal inside the scope of 'between t_1 and t_2', as in (9″):

(9′) There are four persons who are such that they are happy between t_1 and t_2.

(9″) There are four things that are such that they are happy persons between t_1 and t_2.

These are both sensible readings of (9). While (9′) ascribes the sortal 'person' to an ordinary object simpliciter, (9″) ascribes the sortal to an ordinary object at a time. Since (8) and many other cases commit the four-dimensionalist to temporal sorts, and since this commitment renders (9′) unavailable, (9″) is the reading the four-dimensionalist should adopt.

As a last point, it should be mentioned that the widely held thesis that objects do not change with respect to their sorts does not imply that objects fall under their sorts simpliciter—at least not in the strict sense of 'simpliciter' employed here, which indicates absence of all temporal modification.[34] The common thing to say in this case is that what distinguishes unchangeable sorts, such as being a person, from changeable properties, such as being happy or being a teacher, is that, while an object is a person at all times at which it exists, it is a happy teacher only at some of those times. In the remainder of this section, I shall evaluate the prospects of avoiding closure under parthood for both the temporal-sorts view and the atemporal-sorts view.

Sorts and closure

The problem with closure under parthood is that too many things are allowed to have properties at times. If *a* is F at *t*, then everything that overlaps with *a* in its temporal part at *t* is also F at *t*. Since *a* has infinitely many extended temporal parts that overlap with *a* in its instantaneous temporal part at *t*, infinitely many things are F at *t* if *a* is. The claim behind the idea of bringing in ordinary sorts to resolve the difficulties arising from this predicational overkill is that none of the infinitely many proper temporal parts of *a* that are also F at *t* belongs to an ordinary sort; only *a* does. If that is true, then there is reason to contemplate a sortal restriction of ordinary temporal discourse, to the effect that ordinary temporal discourse is true only of ordinary objects, objects belonging to an ordinary sort.

Many temporal predications are already sortally restricted in the simple sense that they contain a sortal term, and hence explicitly invoke a sort. Other temporal predications do not contain a sortal term, and hence do not explicitly invoke a sort. The sample statements raising the problems of cross-temporal counting and continuity were deliberately chosen so as

[34] More precisely, the standard view is that objects have those sorts essentially that are ascribed by substance sortals, such as 'person', but objects may change with respect to sorts that are ascribed by phase sortals, such as 'teacher'. More on this distinction in Sect. 5.6.

not to contain a sortal term. I shall now discuss the question whether predicational overkill still arises for cross-temporal predications that are explicitly sortally restricted. This is the base case for testing whether sortal restriction is able to resolve the problem with closure. If it does not work for the explicitly restricted cases, it does not work at all. I shall focus on the problem of cross-temporal counting, but similar points can be made regarding the problem of continuity. Consider the following cross-temporal predication:

(10) A single person is happy at t_1 and sad at t_2.

This sentence contains the sortal term 'person'. There are two alternative ways for this sortal term to interact with the temporal modifiers 'at t_1' and 'at t_2', which can be made precise in the following two readings of (10):

(11) $\exists!x(\text{Person}(x)\ \&\ \text{At}\ t_1[\text{Happy}(x)]\ \&\ \text{At}\ t_2[\text{Sad}(x)])$
(12) $\exists!x(\text{At}\ t_1[\text{Person}(x)\ \&\ \text{Happy}(x)]\ \&\ \text{At}\ t_2[\text{Person}(x)\ \&\ \text{Sad}(x)])$

where 'Person( )' is the formal equivalent of the ordinary sortal predicate 'is a person'. The difference between (11) and (12) is that in (11) the sortal predicate lies outside the scope of both temporal operators, whereas in (12) the sortal predicate lies inside the scope of each of the temporal operators. In other words, according to (11), something is a person simpliciter, while, according to (12), something is a person at a time. This leads us directly to the distinction between the atemporal-sorts view and the temporal-sorts view.

The atemporal-sorts view says that only maximal unity-interrelated sums of temporal parts can be persons simpliciter. Temporal parts of such sums cannot be persons simpliciter, and hence nothing can be a person at a time. Let us now determine the truth value of (11) and (12) on the assumption of the atemporal-sorts view. Assuming the atemporal-sorts view, it is obvious that (12) cannot be true, since nothing can be a person at a time. (11), on the other hand, can be true. Suppose that Zoe is the single person who is happy at t_1 and sad at t_2. ($\text{TS}_{4\text{D}}$) has the consequence that Zoe-minus, a proper temporal part of Zoe that overlaps with Zoe in its temporal parts at t_1 and at t_2, is also happy at t_1 and sad at t_2. This does not contradict reading (11), however, since Zoe is the only person simpliciter who is happy at t_1 and sad at t_2. For, unlike Zoe, Zoe-minus is not maximal under the unity relation for persons, and hence Zoe-minus is not a person, according to the atemporal-sorts view. This shows that, if the atemporal-sorts view is adopted, then the invocation of sorts does indeed point to a way of dealing with the problem of closure in general. How this might be done will be considered shortly.

The case is different with respect to the temporal-sorts view. The latter says that instantaneous temporal parts are persons simpliciter, and hence that objects having such temporal parts are persons at times. So the temporal-sorts view allows that something be a person simpliciter and that something be a person at a time. Let us now determine the truth value of (11) and (12) on the assumption of the temporal-sorts view. Before doing so, it should be noted that reading (12) is the natural reading for the proponent of the temporal-sorts view to adopt, since it is most plausible to construe all occurrences of sortal predicates in ordinary discourse as being temporally modified—that is, as being governed by temporal operators. The bad news is that neither (11) nor (12) can be true on the temporal-sorts view. According to this view, only instantaneous temporal parts can be persons simpliciter. But, since no instantaneous temporal part can be happy at one time and sad at another, (11) cannot be true, by (TS_{4D}). As regards (12), suppose again that Zoe is the single person who is happy at t_1 and sad at t_2. (TS_{4D}) has the consequence that Zoe-minus, a proper temporal part of Zoe that overlaps with Zoe in its temporal parts at t_1 and at t_2, is also happy at t_1 and sad at t_2. Since these common temporal parts are persons simpliciter, (TS_{4D}) has the further consequence that both Zoe and Zoe-minus are persons at both t_1 and t_2. Since Zoe is distinct from Zoe-minus, (12) cannot be true. This shows that, if the temporal-sorts view is adopted, then the invocation of sorts is of no help whatsoever in dealing with the problem of closure. The problem of cross-temporal counting arises whether sorts are invoked or not. The same can be shown for the problem of cross-temporal continuity. This outcome is very bad news for the four-dimensionalist for two reasons. First, sortal restriction seems to be the only prima facie plausible way of dealing with closure. Secondly, the temporal-sorts view is far superior to the atemporal-sorts view.

Perhaps the four-dimensionalist needs a different kind of restriction. Assuming that the temporal-sorts view is correct, an ordinary sortal term *S* applies to instantaneous stages. Let us call a maximal sum of unity-interrelated *S*-stages a *super-S*. Thus, a super-person is a maximal sum of unity-interrelated person-stages. We saw that the four-dimensionalist loses (12), because sortal restriction is insufficient to block proper temporal parts of a super-person who is a happy person at t_1 and a sad person at t_2 from being happy persons at t_1 and sad persons at t_2 as well. In order to salvage (12), the four-dimensionalist might resort to quantifier domain restriction, claiming that ordinary domains of quantification contain only objects that are maximal under sum unity-relation—that is, typical domains are restricted to super-objects. Since, of all the persons who are happy at t_1 and sad at t_2, only Zoe is a super-person, (12) can be true, assuming that the existential quantifier ranges only over super-objects. The point is that, while sortal restriction has no effect on closure, super-sortal restriction does.

The problem with this reply is that it is ad hoc. A restriction of our quantifiers must be sanctioned by ordinary thought and talk. It is sensible to say that our quantifiers range only over the things that are persons or stones or tables at some time or other. This is a sensible restriction because the sortals 'person', 'stone', and 'table' belong to our everyday conceptual repertoire, and because we typically sort things into persons, stones, tables, etc. Hence, the restriction is based on our ordinary conception of the world. It is not sensible, however, to say that our quantifiers range only over the things that are super-persons or super-stones or super-tables simpliciter. This is not a sensible restriction because the concept of a super-person is a technical label for maximal sums of stages, a label that lacks any basis in ordinary usage; outside the metaphysics room, we do not sort things into super-persons, super-stones, or super-tables. Therefore, the restriction of our quantifiers to super-objects, while satisfying the four-dimensionalist's needs, is entirely unjustified.

One might wonder, in an attempt to soften the threat of predicational overkill, how important a role cross-temporal predications play in ordinary discourse. The answer is that they play a very important role. A cross-temporal uniqueness claim of the sort made by (10) is made by all statements that predicate something of an object x at a time t_1, and that contain a definite description that picks out x at a time t_2 distinct from t_1, such as 'The man who kissed Zoe last night was gone this morning' and 'The currently richest man was poor two years ago'. Far from being rare and special cases, these temporal predications are abundant in ordinary discourse. The fact that they cannot be true on the assumption of the theory of temporal parts, (TS_{4D}), and the temporal-sorts view is therefore a major drawback of the four-dimensionalist outlook.

From instantaneous to extended temporal parts

Faced with this outcome, the four-dimensionalist might try another escape from predicational overkill. We saw that temporal sorts are desirable; ordinary objects are persons or tables at times. By (TS_{4D}), an object a falls under a sort at a time just in case an instantaneous temporal part of a falls under the sort simpliciter. As a consequence, a has many proper temporal parts that are persons at a time. This is why sortal restriction gets no grip on predicational overkill. In the background of this predicament stands the claim, captured in (TS_{4D}), that instantaneous ordinary facts supervene on facts about instantaneous temporal parts. This is the natural four-dimensionalist account of temporal supervenience intuitively motivated by the Flatland/3Dland scenario. Perhaps, however, this view is too narrow. Perhaps ordinary instantaneous facts are determined not only by facts about instantaneous temporal parts but also by facts about extended temporal parts, where an extended temporal part is a sum of non-simultaneous instantaneous temporal parts. A more liberal version of (TS_{4D}) may thus be stated as follows:

(TS'_{4D}) Necessarily, *a* is F at *t* iff *a* has an instantaneous or extended temporal part that is F simpliciter and that occupies a spacetime region that intersects *t*,

where a region *R* intersects *t* just in case *R* has a part that is a member of *t*.[35] This principle allows the combination of the thesis that only maximal unified sums of instantaneous temporal parts fall under ordinary sorts simpliciter, and the thesis that ordinary sorts are instantiated at times. The name 'Zoe' designates a maximal unity-interrelated sum of instantaneous temporal parts. This sum is an extended, improper temporal part of Zoe that is a person simpliciter and that occupies an extended region that intersects various times. By (TS'_{4D}), it follows that Zoe is a person at various times. Given that temporal sorts are admitted and explained in terms of extended temporal parts, sortal restriction is able to block predicational overkill in the basic case (10). We saw that, on the assumption of (TS_{4D}), reading (12) of (10), which is the natural reading, cannot be true, since too many things are persons at t_1 and t_2. This problem with (12) is avoided by (TS'_{4D}), since Zoe is a person at t_1 and at t_2, but none of her proper temporal parts is.

The problem with (TS'_{4D}) is that it is too liberal. In order to understand the problem, let us consider what happens if the idea behind (TS'_{4D}) is applied to the spatial case. Suppose that what determines that *a* has a property in a place *p* is that *a* has a spatial part that occupies a region that intersects *p*. Now suppose that a road has a short stretch in the mountains that is straight. This stretch is part of a bigger stretch of the road that is curved. The curved part occupies a region that intersects the mountain region. By the spatial analogue of (TS'_{4D}), it follows that the road is curved in the mountains. But it is not; the road is straight in the mountains. The conclusion to be drawn is that what the road is like in the mountains is solely determined by what the road's mountain part is like; what bigger parts of the road are like is irrelevant. This spatial case is bound to have temporal analogues. Suppose that Charlie built a house last year. Given that extended temporal parts of Charlie have ordinary properties, it is natural to say that the fact that Charlie built a house over a certain period of time is

[35] I am grateful to Jeremy Butterfield and Nick Shea for discussion of this sort of view.

grounded in the fact that a certain extended temporal part of Charlie that lies in the past builds a house simpliciter. Moreover, there are various past instants, such that Charlie's extended temporal part occupies a region that intersects these instants. By (TS'_{4D}), it follows that Charlie built a house in a instant. But he did not; houses are built over intervals, not at instants. As in the spatial case, the conclusion is that what Charlie is like at an instant is determined solely by what his instantaneous temporal part that lies on this instant is like; what bigger temporal parts of Charlie are and do is irrelevant for truths about an instant.[36]

Another counter-example to (TS'_{4D}) concerns distinctness. By the dyadic version of (TS'_{4D}), *a* is distinct from *b* at *t* iff *a* has an instantaneous or extended temporal part *a** that occupies a region that intersects *t*, and *b* has an instantaneous or extended temporal part *b** that occupies a region that intersects *t*, and *a** is distinct simpliciter from *b**. Notice now that the right-hand side of this biconditional can be satisfied, and hence *a* can be distinct from *b* at *t*, even though '*a*' and '*b*' designate the same object. For one and the same object can have many distinct temporal parts that occupy regions that intersect the same time. Thus, an object can be distinct from itself at a time. This is a highly undesirable result. Similarly to the previous cases, the problem is that (TS'_{4D}) is too liberal about what determines ordinary, instantaneous facts of distinctness by allowing extended temporal parts to do the job. The original version of the four-dimensionalist supervenience principle, (TS_{4D}), avoids this problem. For any instant, an object has many extended temporal parts that intersect that instant, whereas an object has only one instantaneous temporal part that intersects that instant. Since (TS_{4D}) is restricted to instantaneous temporal parts, (TS_{4D}) yields the desired result that no object can be distinct from itself at any time. Thus, to give instantaneous truths an instantaneous supervenience base is the sensible thing to do. The cost, as we saw earlier, is that many cross-temporal predications are lost if the temporal-sorts view is adopted.

The temporal-sorts view says that ordinary sorts apply to instantaneous temporal parts of ordinary objects. By (TS_{4D}), these sorts then apply to ordinary objects at times. The atemporal-sorts view says that ordinary sorts apply to maximal unified sums of instantaneous temporal parts. By (TS_{4D}), these sorts do not apply to ordinary objects at times. That the atemporal-sorts view avoids the problems with (TS'_{4D}), in virtue of embracing (TS_{4D}),

[36] It cannot be true that Charlie built a house at *t*, where *t* is an instant. It can be true, however, that Charlie was building a house at *t*. In the first case the verb is in the simple form, whereas in the second case the verb is in the progressive form. It is therefore important to be aware that the argument above employs only the simple form. We start with the supposition that Charlie built a house at *T*, where *T* is an interval, and conclude by (TS'_{4D}) that Charlie built a house at *t*, without changing the form of the verb.

and that the atemporal-sorts view has a chance of dealing with cross-temporal predications, in virtue of allowing sortal restriction, is little consolation, since the possibility of temporally ascribing ordinary sorts is highly desirable. Nevertheless, it is an interesting question whether the idea of sortal restriction, assuming the atemporal-sorts view, can be worked out into a general solution to the problems of predicational overkill. In the remainder of this section, I shall argue that this cannot be done.

Varieties of sortal restriction

Assuming the atemporal-sorts view, closure under parthood causes no trouble with respect to cross-temporal predications that explicitly invoke a sort. However, the problem of closure was originally stated in terms of temporal discourse that does not explicitly invoke any sort. In order to deal with these cases, it is open to the four-dimensionalist to claim that temporal discourse *implicitly* invokes sorts, that temporal discourse is implicitly sortally restricted. I shall consider two strategies of implementing this idea.

The first strategy is quantifier domain restriction. The idea is to let ordinary domains of quantification be restricted to ordinary objects—that is, to let these domains typically contain only objects falling under ordinary sortal predicates. Notice that this strategy does not avoid closure under parthood. Since (TS_{4D}) is still active, it still holds that, if a is F at t, then every object that overlaps with a in that it has a_t as a temporal part is also F at t. What this strategy does instead is render closure unproblematic in cases involving quantification. As regards the problem of cross-temporal counting, it is now possible for Zoe to be the only one who is happy at t_1 and sad at t_2, because Zoe can be the only object in an ordinary domain that has a happy temporal part at t_1 and a sad temporal part at t_2. That proper temporal parts of Zoe can also be happy at t_1 and sad at t_2 is irrelevant, since such temporal parts are not in the ordinary domain. Compare the spatial case. If a pen is on my desk, then many other things, pen-segments, are also on my desk. But we typically do not care about pen-segments, we care only about pens. By restricting our quantifiers to such things as pens, we can truly say that the pen is the only thing on my desk. As regards the problem of cross-temporal continuity, it is now possible for everything that is happy at t_1 to remain happy at t_2, because it is possible for every object in an ordinary domain that has a happy temporal part at t_1 to have a happy temporal part at t_2 as well.

However, quantifier domain restriction fails to render closure unproblematic in cases involving no quantification. Consider the predicate 'is married'. It is clear that only certain kinds of things can truly be said to be married at a time—namely, persons. If a person P is married at t, then it has a temporal part, p_t, that occupies a member-region of t and that is married. Since P is a sum of temporal parts that is maximal under the unity relation for persons, P has many extended proper temporal parts that also have p_t as a temporal part. By parthood closure, every object that has p_t as a temporal part is married at t. Hence, if P is married at t, then there are many proper temporal parts of P that are also married at t. But proper temporal parts of persons are not themselves persons, according to the atemporal-sorts view, and therefore cannot be married at any time. This is the problem of *predicational fit*. (Note that the intuition that the only things that are married at a time are persons should not be taken as an intuition that the only things that are married simpliciter are persons. We have no intuitions of the latter sort. Ordinary intuitions are intuitions about temporal predications. Atemporal predications are the business of the metaphysician. That a proper temporal part of a person is married simpliciter is therefore not counter-intuitive, but that a proper temporal part of a person is married at a time is counter-intuitive.)[37]

The proponent of quantifier domain restriction might try to avoid the problem of predicational fit by claiming that our intuitions in this case are sortally restricted as well. She might claim that our intuitions in effect say only that no objects in ordinary domains of quantification other than persons can be married at times. This reply overlooks the fact that the statement 'Only a person can be married at a time' is true by definition. For the truth of this statement follows from the definition of 'marriage'—roughly, marriage is a legal union of two persons. And this definition constitutes the unrestricted truth about marriage, not just the truth with respect to ordinary objects. The strategy of sortally restricting our intuitions cannot handle this kind of case.

The second strategy is predicate modification. The four-dimensionalist might claim that a temporal predication of the form '*a* is F at *t*' may implicitly invoke a sort by containing an implicit modifier of the form '*as an S*', yielding '*a* is F *as an S* at *t*', where '*S*' is a sortal term, such as 'person' or 'table'. The sortal information in the elliptical '*a* is F at *t*' may be specified on the basis of the linguistic or the non-linguistic context.[38] Given that ordinary temporal predications may be sortally modified in this way, the four-dimensionalist needs a principle linking sortally relativized

[37] This problem does not arise for the temporal-sorts view. That both Zoe and a proper temporal part of her, Zoe-minus, are married at t is not counter-intuitive because they are both persons at t, in virtue of having the same instantaneous temporal part at t that is a person simpliciter.

[38] I also discuss this form of sortal modification in Sect. 5.6.

temporal facts with spacetime facts. Principle (TS_{4D}) covers the relation of sortally unrelativized temporal facts to spacetime facts. Here is a principle covering the relation of sortally relativized temporal facts to spacetime facts:

(TS^*_{4D}) Necessarily, *a* is F *as an S* at *t* iff *a* is an *S* simpliciter, and *a* has a temporal part that occupies a region that is a member of *t* and that is F simpliciter.

According to this principle, the predicate 'is F' in '*a* is F as an *S* at *t*' is semantically independent of the sortal '*S*'—that is, the extension of 'is F' is not a function of the sortal '*S*'. The predicate 'is F' rather has an unrelativized extension; a temporal part can be F simpliciter. Thus, (TS^*_{4D}) makes '*a* is F as an *S* at *t*' an abbreviation of '*a* is F at *t* and *a* is an *S*'. This has the consequence that the sortal modifier 'as an *S*' is detachable from the left-hand side of (TS^*_{4D}): if *a* is F as an *S* at *t*, then *a* is F at *t*.

Just as quantifier domain restriction does not avoid closure, so the present construal of predicate modification does not avoid closure. Since (TS_{4D}) is still active, it still holds that if *a* is F at *t*, then every object that overlaps with *a* in that it has a_t as a temporal part is also F at *t*. What the strategy of predicate modification does instead is offer a reading of the problematic sentences that renders the latter true. Although the sentence 'Zoe, and only Zoe, is happy at t_1 and sad at t_2' cannot be true if the predicates are read as sortally unrelativized, the sentence can be true if the predicates are read as sortally relativized. Similarly for the sentence 'Everything that is happy at t_1 is still happy at t_2'. However, the present construal of predicate modification, just as quantifier domain restriction, still founders with respect to predicational fit, which concerns the kinds of things that can have a property at a time, as opposed to the number of things. If Zoe is married at *t*, then Zoe-minus is married at *t*, by closure. But Zoe-minus is not a person, on the atemporal-sorts view, only Zoe is. So Zoe-minus cannot be married at any time.

What the four-dimensionalist needs to say in order to avoid the problem of predicational fit is that ordinary temporal discourse cannot be true of proper temporal parts of ordinary objects. Somewhat more precisely, ordinary predicates need to be construed as *irreducibly* sortally modified, so that '*a* is F at *t*' can be true only if it is short for '*a* is F as an *S* at *t*', and the modifier 'as an *S*' cannot be detached. Then a thing that does not fall under a sort *S*, and therefore cannot be F as an *S* at *t*, cannot be F at *t* either. On this construal of sortally relativized predication, Zoe-minus, a proper temporal part of Zoe, cannot be married at a time. For 'Zoe-minus is married at *t*' can be true only if it is short for 'Zoe-minus is married *as a person* at *t*'. But Zoe-minus cannot be married as a person at *t*, since Zoe-minus is not a person. The suggestion that sortal modifiers are irreducible calls for a revision of (TS^*_{4D}) which construes sortal modifiers as reducible. I shall leave the precise nature of the revised principle for later. First, I shall criticize the overall strategy.

Before stating an objection, notice that the above strategy of avoiding predicational fit assumes that the sorts that are implicitly invoked by temporal predications are of the ordinary form, for example, being a person or being a table—that is, sorts that have a corresponding unity relation and for which we have words in ordinary language. In the normal case, proper temporal parts of ordinary objects do not themselves fall under any ordinary sort.[39] But is the restriction to ordinary sorts compulsory? Why cannot temporal predications invoke sorts under which arbitrary temporal parts fall, sorts that lack a corresponding unity relation and for which we lack words in ordinary language, such as being a temporal part of a person?

Here is a good reason why not. Consider again the predicate 'is married'. As stated earlier, it is clear that only certain kinds of objects can truly be said to be married at a time—namely, persons. Taking into account sortal modification, the intuition is that an object can be married only *as a person* at a time. Suppose now that Zoe is married as a person at *t*. Zoe-minus is one of Zoe's proper temporal parts that shares Zoe's temporal part in *t*. Since Zoe-minus is not a person, Zoe-minus is not married as a person at *t*. If 'temporal part of a person' is admitted as a sortal modifier, however, then Zoe-minus is married *as a temporal part of a person* at *t*. This result is counter-intuitive because an object can be married only as a person at a time. To avoid the problem of predicational fit from recurring in this way, sortal modification of temporal predications must be restricted to ordinary sortals, none of which applies to proper temporal parts of persons.

The proposal under consideration is to block ordinary temporal discourse from saying anything true about objects that do not fall under any ordinary sort. I shall now argue that this is much too strong a restriction. Consider an object's spatial parts. A table, for instance, has various spatial parts at any time at which it exists. The spatial parts of a table are not themselves tables, just as the temporal parts of a table are not themselves tables. Some spatial parts of a table belong to different sorts, such as a table's legs. But these 'ordinary' parts do not exhaust the table's spatial parts. According to the

[39] A case in which an object that falls under an ordinary sort has a proper temporal part that itself falls under an ordinary sort is a case in which two distinct ordinary objects coincide. In the case in which a lump of clay is formed into a statue, to mention one example, there is a lump of clay and a statue, and the statue is a proper temporal part of the lump of clay. Cases of coincidence are special cases and will be discussed in Sect. 5.6.

doctrine of arbitrary spatial parts, for any way of dividing the spatial region occupied by a table at a time t into subregions, the table has corresponding spatial parts at t that occupy these subregions at t. If the doctrine of arbitrary spatial parts is true, then a table has spatial parts that, unlike its legs, do not belong to any sort,—for instance, the left half of a table. However, if temporal predication is sortally modified in the way currently under consideration, spatial parts that do not fall under any sort cannot have any properties at any times. The left half of a table fails to have a shape at any time, because it is not the case that the left half of a table has a shape *as an S* at any time, for any sortal term '*S*'. That arbitrary spatial parts cannot be the subject of temporal discourse is problematic because these sortless spatial parts must somehow be individuated and therefore must have various properties at various times, such as a certain shape and mass. Hence, the thesis that temporal predication is sortally modified cannot be held in combination with the doctrine of arbitrary spatial parts.

In response, one might point out that in this argument only ordinary sorts are admitted. In order to save the doctrine of arbitrary spatial parts, one might opt for a less restrictive proposal and say that, while it is true that the left half of a table does not fall under any ordinary sort, it does fall under such a non-ordinary sort as being a spatial part of a table. This is not a sensible move for the four-dimensionalist to make, though. As we saw above, the strategy of sortal modification is plausible only if restricted to ordinary sortals, and hence only if sortals such as 'temporal part of a table' are ruled out. As a consequence, sortals such as 'spatial part of a table' must be ruled out as well. As an alternative response, one might reject the doctrine of arbitrary spatial parts and try to construe ordinary talk of such things as table-halves as non-literal. This second move, like the first one, is unavailable to the four-dimensionalist. For the doctrine of arbitrary temporal parts, which lies at the heart of the theory of temporal parts, mirrors the doctrine of arbitrary spatial parts. What is more, the doctrine of arbitrary temporal parts and the doctrine of arbitrary spatial parts are, from the four-dimensionalist perspective, most plausibly viewed as restricted versions of the generalized doctrine of spatiotemporal parts (see Section 4.3). Since this space–time symmetry is constitutive of the four-dimensionalist universe, the four-dimensionalist had better keep arbitrary spatial parts and drop sortal modification.

I said that the construal of sortal modifiers as irreducible requires a revision of (TS^*_{4D}). How should this principle be revised? If a sortal modifier is semantically irreducible, then its predicate is semantically dependent on it. If a is F as an S at t, then 'is F' is semantically dependent on 'S'; the predicate has an extension only relative to the sort being an S (see Section 5.6 for details). The principle linking irreducibly sortally relativized temporal facts with spacetime facts looks as follows:

(TS^{**}_{4D}) Necessarily, *a* is F *as an S* at *t* iff *a* has a temporal part that occupies a region that is a member of *t*, and that is F *as an S* simpliciter.

Since the predicate 'is F' is semantically dependent on the sortal modifier, it is accompanied by this modifier on the right-hand side as well as on the left-hand side. Notice now that, if something is F as an *S*, it does not follow that it is F, but it clearly follows that it is an *S*. If something is happy as a person, then it is obviously a person. According to (TS^{**}_{4D}), if Zoe is happy as a person now, then Zoe has an instantaneous temporal part now that is happy as a person simpliciter, and hence Zoe has an instantaneous temporal part now that is a person simpliciter. What this shows is that the construal of sortal modifiers as semantically irreducible requires the temporal-sorts view, according to which ordinary sorts apply to temporally unextended stages, as opposed to temporally extended sums of stages. Since we already know that four-dimensionalism in combination with the temporal-sorts view is threatened by cross-temporal predications that explicitly invoke a sort, it is no surprise that cross-temporal predications that implicitly invoke a sort pose an equal threat. Thus, 'Zoe, and only Zoe, is happy as a person at t_1 and sad as a person at t_2' cannot be true on the assumption of (TS^{**}_{4D}), because each of Zoe's extended temporal parts that shares Zoe's temporal parts at t_1 and t_2 is also happy as a person at t_1 and sad as a person at t_2.

To conclude, the strategy of sortally restricting temporal discourse—either by way of quantifier domain restriction or by way of predicate modification—is implausible in the four-dimensionalist framework. The threat arising from closure under parthood—the threat of predicational overkill—therefore remains a serious one whether the four-dimensionalist adopts the temporal-sorts view or the atemporal-sorts view. I take this to be a major flaw of the temporal-parts account of temporal supervenience.

4.6 THE TEMPORAL-COUNTERPARTS ACCOUNT

The theory of temporal parts has a close variant: the theory of temporal counterparts. Both theories adopt an ontology of temporally extended spacetime worms with temporal parts but differ concerning what counts as an ordinary object—that is, they differ regarding the things we refer to in ordinary discourse. According to the theory of temporal parts, a spacetime worm counts as an ordinary object, but, according to the theory of temporal counterparts, each instantaneous temporal part of a spacetime worm counts as an ordinary object, and all the objects that are temporal parts of a spacetime worm are temporal counterparts of each other and of themselves.[40] I shall conclude this chapter with a discussion of an account of temporal supervenience based on the theory of temporal counterparts, comparing the temporal-counterparts account to the temporal-parts account and arguing that the former is even worse off than the latter.

Temporal counterparts and supervenience

The theory of temporal counterparts has the consequence that an ordinary object, being an instantaneous temporal part of an extended spacetime worm, occupies a unique temporally unextended spacetime region. This thesis, which was earlier labelled four-dimensionalism*, represents the middle ground between three-dimensionalism and four-dimensionalism:

(4D*) (i) an ordinary object occupies a unique spacetime region, and (ii) this spacetime region is temporally unextended.

An ordinary object that is spatiotemporally located in this way has temporal counterparts that are themselves ordinary objects. How, then, are ordinary properties, such as shapes, instantiated in spacetime? Recall that the friend of temporal parts may say that shapes are instantiated simpliciter by the temporal parts of an ordinary object. The friend of temporal counterparts, by contrast, may say that shapes are instantiated simpliciter by ordinary objects themselves and their counterparts. So we have counterpart-theoretic accounts of spatiotemporal location and instantiation. These accounts yield two supervenience theses: (*a*) the facts of persistence logically supervene on facts about the spatiotemporal location of objects and their temporal counterparts; and (*b*) the facts of temporal instantiation logically supervene on facts about the atemporal instantiation of properties by objects and their temporal counterparts. The remaining task is to explain these supervenience theses.

The explanation of (*a*) and (*b*) requires, first, an account of temporal predication. I suggest that the best bet in the case of temporal counterparts, as in the case of temporal parts, is the representational account of temporal predication (set out in detail in Section 4.1). This account construes 'at *t*' in '*a* is F at *t*' as a sentence modifier to be interpreted as 'according to *t*', where '*a* is F according to *t*' is defined as '*a* has a representative in *t* that is F simpliciter'. This much is familiar. The next task is to give an account of temporal representation, an account of what *a*'s representative in *t* is. Here the theory of temporal counterparts differs from the theory of

[40] For a more formal statement of the theory of temporal counterparts, see Sect. 2.2.

temporal parts, because the former should construe temporal representation as follows:

(REP_{4D^*}) Necessarily, *a* has a representative in *t* that is F iff *a* has a temporal counterpart that is F and that occupies a region that is a member of *t*.

Combining the representational account of temporal predication with the counterparts account of temporal representation yields the principle that constitutes the desired bridge from ordinary temporal facts to counterpart facts:

(TS_{4D^*}) Necessarily, *a* is F at *t* iff *a* has a temporal counterpart that is F and that occupies a region that is a member of *t*.

Given (TS_{4D^*}), the claims of the theory of temporal counterparts—that an object and its temporal counterparts each occupies an instantaneous spacetime region and that each has various properties simpliciter—entail that an object persists and has various properties at various times. This is the temporal-counterparts account of persistence supervenience and temporal-instantiation supervenience.

Intrinsicness lost

While overall very similar to the temporal-parts account of temporal supervenience, the temporal-counterparts account is different in an important respect. According to the temporal-counterparts account, how an object is in ordinary time does not supervene on how that object is in spacetime. (4D*) says that an ordinary object *a* occupies no more than a single instantaneous spacetime region. By (TS_{4D^*}), this entails merely that *a* exists at a single time. It does not entail that *a* exists at various times, that *a* persists. Instead, the persistence of *a* supervenes on the spatiotemporal location of *a* and the spatiotemporal location of temporal counterparts of *a*. This feature is worrying, since the persistence of *a* through ordinary time is intrinsic in the sense that *a* could be the only object in ordinary space and time and still persist. The temporal-counterparts account is unable to capture the intrinsicness of persistence at the level of spacetime. The persistence of *a* supervenes on extrinsic features of *a*; it supervenes on facts about *a* and about objects wholly distinct from *a*. So an ordinary object cannot be the sole occupant of spacetime and persist through ordinary time. The temporal-parts account does not have this consequence, since how an object is in ordinary time supervenes on how that object is in spacetime, irrespective of any other, wholly distinct objects. (4D) says that an ordinary object *a* occupies a unique spacetime region that is temporally extended. By the theory of temporal parts and (TS_{4D}), this entails that a exists at various times, and hence that a persists. The temporal-parts account is thus able to capture the intrinsicness of persistence at the level of spacetime. This feature puts the temporal-parts account ahead of the temporal-counterparts account. With taking ordinary time seriously comes the imperative to try to capture as many features of ordinary time at the level of spacetime as possible. Accordingly, the intrinsicness of persistence is worth preserving. While the failure to preserve this intrinsicness might not constitute a direct objection to the temporal-counterparts account, it does constitute a reason in favour of the temporal-parts account.

Closure under counterparthood

A further consequence of the temporal-counterparts account is that temporal predication is *closed under the temporal-counterpart relation* in the following sense. Since all that is required to be F at t is to have a temporal counterpart, a_t, that is located at t and that is F, every object that has a_t as a temporal counterpart is F at t. Under normal circumstances, the counterpart relation is reflexive, symmetric, and transitive.[41] So, if a has a_t as a counterpart, then each of a's counterparts, including a_t itself, has a_t as a counterpart. Hence, if a is F at t, then every temporal counterpart of a is also F at t.

We saw that closure under parthood leads to predicational overkill. Does the same hold for closure under counterparthood? Let us begin with the problem of cross-temporal continuity. The temporal-parts account renders (7) necessarily false because not every object that has a happy temporal part at t_1 can have a happy temporal part at t_2. An instantaneous temporal part that has itself as a happy temporal part at t_1 is an example. The temporal-counterparts account, on the other hand, has no trouble with (7), because it is possible that every object that has a happy temporal counterpart at t_1 also has a happy temporal counterpart at t_2.

To see where counterpart closure goes wrong, reconsider example (4). The latter is regimented as (5). By the temporal-counterparts account, if Zoe is happy at t_1 and sad at t_2, then Zoe has a temporal counterpart, z_{t_1}, that is located at t_1 and that is happy, and Zoe has a temporal counterpart, z_{t_2}, that is located at t_2 and that is sad. Since the counterpart relation is reflexive and transitive, both z_{t_1} and z_{t_2} have a counterpart at t_1 that is happy and one at t_2 that is sad, and so both z_{t_1} and z_{t_2} are happy at t_1 and

[41] In Parfit-style cases of fission, the temporal-counterpart relation is non-transitive. See Parfit (1975) and Sider (1996).

sad at t_2. But z_{t_1} and z_{t_2} are distinct simpliciter. Hence, if Zoe is happy at t_1 and sad at t_2, then Zoe is not the only thing that is happy at t_1 and sad at t_2, which makes (4) impossible. The problem of cross-temporal counting thus remains.[42]

Can closure under counterparthood be blocked by sortal restriction? Notice first that sorts in the theory of temporal counterparts are treated as sorts are treated by the temporal-sorts variant of the theory of temporal parts, which was earlier shown to be superior to the atemporal-sorts variant of the latter. According to both theories, the things that fall under ordinary sorts simpliciter are instantaneous things. The theories differ concerning which things we refer to and quantify over in ordinary discourse. On the temporal-sorts version of the theory of temporal parts, we refer to and quantify over spacetime worms, sums of temporal parts, each temporal part belonging to an ordinary sort simpliciter, whereas on the theory of temporal counterparts we refer to and quantify over temporal parts of spacetime worms, each temporal part, or temporal counterpart, belonging to an ordinary sort simpliciter. We have seen that sortal restriction of temporal discourse, whether explicit or implicit, has no effect on the problem of cross-temporal counting if the temporal-parts account is combined with temporal sorts. For analogous reasons, sortal restriction of temporal discourse, whether explicit or implicit, has no effect on the difficulties arising from closure under counterparthood. If a person is happy at t_1 and sad at t_2, then it has many distinct temporal counterparts that are also happy at t_1 and sad at t_2, and that are also persons at t_1 and t_2. Thus, Zoe cannot be the only person who is happy at t_1 and sad at t_2.

The cases that create trouble for the theory of temporal counterparts are ones in which we speak about what happens to a particular thing over a certain period of time. Sider, a recent advocate of the theory of temporal counterparts—or the *stage view*, as he calls it—notices the difficulty pointing out that the theory has a hard time giving the intuitively correct answer to such questions as 'How many people have been sitting in my office during the last hour?'[43] In response, Sider suggests combining the theory of temporal counterparts with the theory of temporal parts. The

[42] Analogously to the case of the temporal-parts account, the problem of cross-temporal counting is fatal for the temporal-counterparts account only with respect to examples such as (4), but not with respect to simpler examples such as (1). The proponent of temporal counterparts may avoid the trouble with (1) by saying that, if we count by temporally unrelativized identity, identity simpliciter, many temporal counterparts of Zoe are happy at t if Zoe is happy at t. But if we count by temporally relativized identity, identity at a time, then Zoe is the only one who is happy at t, because many temporal counterparts that are distinct simpliciter can be identical at t.

[43] See Sider (1996: 448).

difference between the two theories, so says Sider, may be understood as a semantic one, as a difference concerning what to call 'persons'. According to Sider, the sortal term 'person' is ambiguous. The things that typically fall in the extension of 'person' are temporal counterparts. But in certain special cases the things that fall in the extension of 'person' are maximal sums of temporal counterparts. Correspondingly, on one disambiguation of the sortal 'person', the truth conditions of a temporal predication containing 'person' involve temporal counterparts, and on the other disambiguation the truth conditions involve temporal parts. Among the special cases, so says Sider, are those in which we trace a particular individual through time—cases that I have called 'cross-temporal predications'. Thus, when we say that a single person is happy at t_1 and sad at t_2 or that a single person has been sitting in my office during the last hour, 'person' is used to range over sums of temporal counterparts, instead of temporal counterparts. As Sider admits, the ambiguity strategy is somewhat ad hoc and uneven. But he claims that the price is right, and so these drawbacks may be tolerated.

The first thing to notice about this strategy is that it presupposes the atemporal-sorts version of the temporal-parts account, according to which spacetime worms are persons. The difference that Sider describes disappears, however, if the more plausible temporal-sorts version is presupposed, which calls exactly those things 'persons' that the temporal-counterparts account calls 'persons'. The desired ambiguity should therefore be located not in the extension of 'person' or 'chair', but rather in what we ordinarily designate by the use of names: the name '*a*' in '*a* is F at *t*' usually designates an instantaneous temporal part of a spacetime worm, a stage; but in some special cases, such as cross-temporal predications, '*a*' designates a spacetime worm instead. In other words, the ambiguity should be located in what we ordinarily describe as being a person *at a time*.

Secondly, and more importantly, assuming the temporal-sorts version of the temporal-parts account, Sider's ambiguity strategy fails to take care of the problem of cross-temporal counting for the temporal-counterparts account, because the temporal-parts account stumbles over exactly the same difficulty. The only way to deal with the problem of cross-temporal counting is to adopt the atemporal-sorts version of the temporal-parts account and to claim that temporal predication is implicitly, if not explicitly, sortally modified. However, the combination of the thesis that temporal predications are sortally modified with the atemporal-sorts version of the temporal-parts account was earlier shown to be untenable for other reasons.

To sum up, just as the temporal-parts account of temporal supervenience is closed under parthood, so the temporal-counterparts account is closed under counterparthood. And, like closure under parthood, closure under counterparthood leads to predicational overkill. Since it is unclear how the trouble with closure under counterparthood could be avoided, I conclude that the temporal-counterparts account is no better off than the temporal-parts account.

5

Three-Dimensionalist Supervenience

The four-dimensionalist temporal-parts account of temporal supervenience has a three-dimensionalist rival that shares the temporal-parts account's main virtues and avoids its main drawbacks: the temporal-regions account. So I shall argue in the present chapter. The statement of the temporal-regions account will be followed by a comparison with the temporal-parts account. The remainder of the chapter will be concerned with various consequences and apparent difficulties of three-dimensionalist supervenience.

5.1 THE TEMPORAL-REGIONS ACCOUNT

In this section, I will develop the temporal-regions account of temporal supervenience in three steps. The first step is a semantic account of temporal predication. The second step is a metaphysical account of spatiotemporal location and spatiotemporal instantiation. The third step is the construction of a bridge principle that links ordinary temporal facts with spatiotemporal facts.

The representational account of temporal predication

The first step in developing the temporal-regions account is to give an account of temporal predication. This step is easy in the present case, since the temporal-regions account shares with the temporal-parts account the representational account of temporal predication, which was stated in detail in Section 4.1. I shall briefly review the main theses.

An account of temporal predication is a specification of the logical and semantic function of the temporal modifier 'at *t*' in temporal predications with the surface form '*a* is F at *t*'. According to the representational account of temporal predication, the temporal modifier 'at *t*' appears at the level of logical form as a sentential temporal operator 'At *t*': At *t*[F(*a*)]. The operator 'At *t*' is semantically reducible in that it can be given the representational interpretation 'Ac t', to be read as 'according to t', and 'Ac t' can be given the following definition:

(Def2) Ac $t[F(a)] =_{df}$ t represents a as being F.

The right-hand side of (Def2) may be further analysed in terms of the notion of a *representative*:

(Def3) x represents a as being F $=_{df}$ a has a representative in x that is F simpliciter.

If 'At $t[F(a)]$' is interpreted as 'Ac $t[F(a)]$', and, if (Def2) and (Def3) are put together, then the right-hand side of (Def3) with 'x' replaced by 't' gives the meaning of 'At $t[F(a)]$'. Since the right-hand side of (Def3) is an atemporal predication, monadic temporal predications may be given temporally unmodified truth conditions:

(T_2) 'At $t[F(a)]$' is true $\equiv \exists x(\mathrm{Rep}(x, a, t)\ \&\ F(x))$

Following analogous steps, a similar semantic reduction may be achieved for dyadic temporal predications:

(T'_2) 'At $t[R(a, b)]$' is true $\equiv \exists x \exists y(\mathrm{Rep}(x, a, t)\ \&\ \mathrm{Rep}(y, b, t)\ \&\ R(x, y))$

This is the representational account of temporal predication. In Section 4.1, this account was shown to avoid various difficulties for the relational account and the intensional account of Chapter 3. This superiority of the representational account, together with the fact that the main problems with four-dimensionalist supervenience are independent of this account, make the latter a viable choice for a three-dimensionalist approach to temporal supervenience.

Three-dimensionalism and regional instantiation

The second step in explaining temporal supervenience is an account of spatiotemporal location and of spatiotemporal instantiation. The task is to specify a spatiotemporal supervenience base for ordinary facts of temporal existence and temporal instantiation. Starting with persistence, the supervenience base is specified by three-dimensionalism or four-dimensionalism, which are the two main accounts of spatiotemporal location stated in Chapter 2. In Chapter 4 we considered perdurantism, the thesis that persistence logically supervenes on four-dimensionalist facts about the spatiotemporal location of temporal parts of objects. According to perdurantism, the ordinary persistence fact that a exists at different times is entailed by the spatiotemporal fact that a has different, instantaneous temporal parts. The thesis under consideration now is endurantism, the thesis that the facts of persistence logically supervene on the following three-dimensionalist facts about spatiotemporal location:

(3D) (i) an ordinary object occupies multiple spacetime regions, and (ii) these spacetime regions are temporally unextended, or instantaneous, and non-simultaneous.

Stating endurantism is one thing. Explaining how an object's occupying multiple instantaneous regions determines its persistence is another. This explanatory task will be taken up in the following subsection.

Given endurantism, what spacetime facts do ordinary facts of temporal instantiation supervene on? This question is essentially the problem of spatiotemporal instantiation: how are ordinary properties, such as shapes and colours, instantiated in spacetime? The three-dimensionalist has several options here, two of which we have already encountered. First remember that there are no times according to the spacetime conception, and that therefore the ordinary notion of having a shape at a time has no application to this conception. Given that shapes cannot be had relative to a time, which component of three-dimensionalist occupied spacetime has shapes and how? The three-dimensionalist has the option of construing shapes as some sort of relation—an option familiar from the relational account of temporal predication—or as properties the instantiation of which is somehow relativized—an option familiar from the intensional account. As we saw earlier, the unrelativized having of shape-relations and the relativized having of shape-properties are independent of any particular mode of spatiotemporal location. That an ordinary object has a shape-relation to a spacetime region or to a set of spacetime regions, or that an object has a shape-property relative to a spacetime region or relative to a set of spacetime regions, does not imply anything about how the object is located in spacetime, and indeed does not even imply that the object is spatiotemporally located at all. This is why these accounts of spatiotemporal instantiation are compatible with any account of an object's spatiotemporal location, and hence why each of them is compatible with both three-dimensionalism and four-dimensionalism. I mention these two construals of spatiotemporal instantiation merely as a matter of completeness, since they have already been discussed and rejected in Chapter 3. Moreover, these construals are incompatible with the representational account of temporal predication. According to the latter, shapes are properties instantiated simpliciter, as opposed to properties instantiated relative to a time or relations.

The three-dimensionalist holds that an ordinary object occupies many instantaneous spacetime regions. The simplest way of painting shapes and masses and colours into this picture is to say that the object that occupies the spacetime regions has shapes and masses and colours simpliciter. The resulting picture is, of course, much too simple, since the idea that *a* has a bent shape simpliciter is outright incompatible with the ordinary fact that *a* can change its shape to the effect that it is bent-shaped at one time and straight-shaped at another. For, given that the fact of *a*'s being bent-shaped at t_1 is grounded in the fact that *a* occupies a certain instantaneous region and is bent-shaped simpliciter, and that the fact of *a*'s being straight-shaped at t_2 is grounded in the fact that *a* occupies another region and is straight-shaped simpliciter, it follows that *a* is bent-shaped and straight-shaped simpliciter, which is impossible. This threat of contradiction is what sets off the metaphysical problem of change. What alternative account of spatiotemporal instantiation is there, given that three-dimensionalism denies an ordinary object temporal parts to which ordinary properties could be relegated? The answer I propose is that shapes and masses and colours are properties had simpliciter by instantaneous spacetime regions occupied by ordinary objects. Ordinary properties are instantiated neither by objects nor by temporal parts of objects. They are instantiated rather by the regions ordinary objects occupy. Spatiotemporal instantiation is *regional instantiation*. Given this account of spatiotemporal instantiation, the three-dimensionalist may say that the ordinary fact that *a* is F at *t* is entailed by the spacetime fact that *a* occupies a certain instantaneous spacetime region that is F simpliciter. Moreover, the threat of contradiction that drives the metaphysical problem of change is banned by letting the ordinary fact that *a* changes be entailed by the fact that *a* occupies different regions that have different properties simpliciter.

The account of spatiotemporal instantiation as regional instantiation raises a question of generality. Could the three-dimensionalist characterize the supervenience base of the instantiation of temporary, changeable, properties in the way just sketched, while adopting a different account of the supervenience base of the instantiation of permanent, unchangeable, properties? For example, could the three-dimensionalist say that the fact that an object instantiates a temporary property at a time is grounded in the fact that the object occupies a spacetime region that has the property simpliciter, whereas the fact that an object instantiates a permanent property at a time is grounded in the fact that the object itself instantiates this property simpliciter? As we saw in Section 4.1 when discussing a similar question of generality addressed to the four-dimensionalist, to specify different types of supervenience base for facts of temporal instantiation requires different semantic treatments of 'at *t*' in '*a* is F at *t*', depending on whether 'is F' is a temporary or a permanent predicate of *a*. The postulation of an ambiguity in 'at *t*' is unjustified, however, because the question how 'at *t*' should be treated semantically is independent of the question whether *a* is F at all times of its existence or only at some of those times. If we assume that the three-dimensionalist aims for a unitary treatment of temporal modification, she is committed to the account of spatiotemporal instantiation as regional instantiation in full generality.

As in the case of persistence supervenience, it is not enough just to state the thesis that regional instantiation underlies ordinary temporal instantiation. The thesis requires explanation, for three-dimensionalist supervenience is not merely a brute fact. Before proceeding to the required explanation, a positive feature of this account of the supervenience base of temporal instantiation should be noted. In Chapter 3 I put forward a constraint on any account of the supervenience base of temporal instantiation. It seems undeniable that, if an object has a property at a time, then the object must exist at a time. Given the assumption that the facts of ordinary time and their relations logically supervene on facts about spacetime and their relations, this requirement should translate to the level of spacetime, yielding the requirement that the facts of spatiotemporal instantiation underlying ordinary facts of temporal instantiation imply certain facts of spatiotemporal location underlying ordinary facts of temporal existence. This is the temporal-existence requirement. It is a constraint on the relation between the supervenience base of temporal instantiation and the supervenience base of temporal existence. As was shown in Chapter 3, attempts to specify the supervenience base of having a shape at a time in terms of shape-relations or relativized shape-properties fail the temporal-existence requirement because of the compatibility of an object's having shape-relations and its having relativized shape-properties with the possibility that the object may not be spatiotemporally located at all. An account of the supervenience base of having a shape at a time in terms of a temporal part of an object having a shape simpliciter and occupying a member-region of this time, on the other hand, straightforwardly captures the temporal-existence requirement, as was shown in Section 4.1. The present account of the supervenience base of temporal instantiation in terms of regional instantiation shares this positive feature with the temporal-parts account. The task is to specify a fact A of spatiotemporal instantiation that entails that *a* has a property at a time and a fact B of spatiotemporal location that entails that *a* exists at a time, such that A implies B. According to the present account, the fact that *a* has a certain shape at a time *t* is entailed by the fact that *a* occupies a spacetime region that has a certain shape simpliciter and that is a member of *t*. Moreover, the fact that *a* exists at *t* is entailed by the fact that *a* occupies a region that is a member of *t*. Since the fact that *a* occupies a region that has a certain shape and is a member of *t* trivially implies that *a* occupies a region that is a member of *t*, the temporal-existence requirement is satisfied.

Putting the pieces together: The temporal-regions account of temporal supervenience

Our three-dimensionalist puts forward the following two supervenience theses: (*a*) the facts of persistence logically supervene on facts about the spatiotemporal location of objects; and (*b*) the facts of temporal instantiation logically supervene on facts about the atemporal instantiation of properties by instantaneous spacetime regions occupied by objects. The remaining task is to *explain* persistence supervenience and temporal-instantiation supervenience as construed by (*a*) and (*b*).

In order to do so, let us return to the representational account of temporal predication. This account with its semantic interpretation of temporal modifiers as representational modifiers yields the following reductive equivalence:

(RED) $\Box[\text{At } t[\text{F}(a)] \equiv \exists x(\text{Rep}(x, a, t) \,\&\, \text{F}(x))]$

This principle construes representation of *a* by *t* in terms of a representative of *a* in *t*. What is a representative of *a* in *t* within the three-dimensionalist framework? Answering this question will yield a three-dimensionalist explanation of temporal supervenience.

In the previous chapter we considered a four-dimensionalist account of temporal representatives, according to which a temporal representative of an ordinary object *a* is a temporal part of *a*. For such a representative to be 'in' a time *t* is for the temporal part to occupy a spacetime region that is a member of the set of spacetime regions that is *t*. The three-dimensionalist may now offer the following alternative account of temporal representatives: a temporal representative of an ordinary object is an instantaneous spacetime region that the object occupies—that is, the representative-relation is the occupation-relation. Since the three-dimensionalist holds that an ordinary object occupies many instantaneous spacetime regions, an object has many temporal representatives. For such a temporal representative to be 'in' a time *t* is for the instantaneous spacetime region to be a member of the set of regions that is *t*. Letting 'IR(*x*)' mean that *x* is an instantaneous region, this three-dimensionalist account of temporal representatives may be stated as follows:

($\text{REP}_{3\text{D}}$) $\Box[\text{Rep}(x, a, t) \equiv \text{IR}(x) \,\&\, \text{O}(a, x) \,\&\, x \in t]$

Given this account of temporal representatives, temporal supervenience may be explained. The three-dimensionalist account of temporal representation functions as a bridge principle linking the three-dimensionalist's facts of spatiotemporal location and instantiation with the ordinary facts of temporal existence and instantiation via the representational account of temporal predication. Beginning with the sentence-modifier reading of '*a* is F at *t*', 'At *t*[F(*a*)]', then interpreting 'At *t*' as representational, and adding (RED) and (REP_{3D}), yields the following biconditional, which forms the backbone of the three-dimensionalist temporal-regions account of temporal supervenience:

(TS_{3D}) $\Box[\text{At } t[F(a)] \equiv \exists R(\text{IR}(R) \,\&\, \text{O}(a, R) \,\&\, R \in t \,\&\, F(R))]$
Necessarily, *a* is F at *t* iff *a* occupies an instantaneous spacetime region *R*, such that *R* is a member of *t* and *R* is F simpliciter.

Now recall the three-dimensionalist supervenience theses (*a*) and (*b*): the facts of temporal persistence logically supervene on facts about the spatiotemporal location of objects; and the facts of temporal instantiation logically supervene on facts about the atemporal instantiation of properties by instantaneous spacetime regions occupied by objects. Theses (*a*) and (*b*) are explained by (TS_{3D}) in virtue of being consequences of this principle. Thesis (*b*) is an obvious consequence of (TS_{3D}). To see that (*a*) is also a consequence of (TS_{3D}), apply the latter to an ordinary existence statement '*a* exists at *t*': *a* exists at *t* iff *a* occupies an instantaneous region that is a member of *t* and that exists simpliciter. This may be simplified: *a* exists at *t* iff *a* occupies an instantaneous region that is a member of *t*. Hence facts about the spatiotemporal location of an object entail facts about the object's temporal existence, which is what thesis (*a*) asserts. By constructing (TS_{3D}) out of the representational account of temporal predication and the temporal-regions account of temporal representation, the three-dimensionalist obtains a smooth explanation of persistence supervenience and temporal-instantiation supervenience.

5.2 THE STATUS OF REGIONAL INSTANTIATION

These are the basics of the temporal-regions account of temporal supervenience. The most notable feature of this account is its construal of spatiotemporal instantiation in terms of spacetime regions having ordinary properties simpliciter. In this section I shall make a few remarks on the intuitive and theoretical status of regional instantiation.

As regards the intuitive status, regional instantiation raises the immediate concern that the idea that a spacetime region can be happy or married to another region constitutes a violation of common sense, for we ordinarily think that a person is happy and a couple is married, not the regions occupied by the person or the couple. In order to evaluate this objection, it is important to distinguish the claim that a spacetime region is happy or married at a time from the claim that a region is happy or married simpliciter. To have an attribute at a time is to have the attribute in the ordinary sense. To have an attribute simpliciter is to have the attribute in a different, more technical sense. According to regional instantiation, a spacetime region is happy or married simpliciter. By (TS_{3D}), something can have a property at a time only if it occupies a certain region that has this property simpliciter. Since spacetime regions do not occupy themselves or other regions, regions do not have any properties or relations at any time. It is rather an ordinary object occupying a spacetime region that is happy or married at a time.[1] So regional instantiation does not go against what we ordinarily think. Ordinary thought and talk are confined to ordinary time. Since all our intuitions are temporally modified, and since regional instantiation is a doctrine concerning atemporal predications only, a doctrine that does not translate to the ordinary level, the question of the intuitive status of regional instantiation does not arise. For the question as to what, if anything, is happy or married simpliciter lies outside the scope of ordinary intuition. It is up to the metaphysician to describe the world in technical, temporally unmodified language as containing whichever things related in whichever way suitable to make sense of its ordinary description in temporally modified language. To this end, the proponent of regional instantiation describes spacetime regions as being happy or married simpliciter. This strategy is not obviously counter-intuitive, for it does not replace ordinary talk by region-talk. Spacetime regions operate exclusively below the surface.

To get a better intuitive grip on the distinction between '*a* is F simpliciter' and '*a* is F at a time', I propose to think of instantaneous spacetime regions as *temporal states*—the instantaneous version of temporally extended events. A region that is happy simpliciter is an instantaneous state of happiness. Moreover, to occupy a region that is happy simpliciter is to be in a state of happiness. The fact that Zoe is happy at a time is thus entailed by the fact that Zoe is in a state of happiness that obtains at some time. Similarly, to say that region R_1 is married simpliciter to region R_2 is to say that R_1 and R_2 form an instantaneous state of being married. And to occupy a pair of regions that is married simpliciter is to be in a state of being married. The fact that Suzie and Charlie are married at a time is thus entailed by the fact that Suzie and Charlie are in a state of being married that

[1] If spacetime regions do not occupy themselves or other regions, then spatial regions, ordinary places, do not occupy themselves or other places either. Then how can something occupy a place at a time, given the dyadic version of (TS_{3D})? The answer was given in Sect. 4.3. The sentence '*a* occupies *p* at *t*', as the three-dimensionalist reads it, is not a temporally modified predication to which (TS_{3D}) applies. The sentence rather has the form 'O(*a*, AT(*p*, *t*))', where 'AT(*p*, *t*)' is a singular term designating the intersection of *p* and *t*.

obtains at some time. What the construal of instantaneous regions as states illustrates is that the temporal-regions account is far from claiming that the roles played by ordinary objects at the level of ordinary time are played by instantaneous regions at the level of spacetime. Being happy simpliciter is not the spacetime equivalent of being happy at a time. Rather, being in a temporal state of happiness is the spacetime equivalent of being happy at a time. This talk of temporal states is not meant to play any technical role. State-talk is rather intended as a device to render regional instantiation intuitively more accessible.[2]

As regards the theoretical status of regional instantiation, the first point to notice is that regional instantiation does not stand in conflict with the ontology of spacetime, as understood here. Following standard practice, spacetime points and regions were introduced as ontological primitives. Questions concerning the metaphysical nature of spacetime points and regions—are they *sui generis* kinds of entities or are they constructed from more basic kinds of entities?—are left entirely open. Points and regions are characterized functionally, in terms of certain theoretical roles defined by modern physics, in particular by relativity theory. To these physical roles we added certain metaphysical roles that spacetime points and regions are meant to play: the role of being occupied by ordinary objects and the role of instantiating ordinary properties simpliciter. As in the case of the physical roles, these metaphysical roles are not constrained by what spacetime points and regions are. There is no question as to whether spacetime points and regions are the kinds of things that can play these roles, since spacetime points and regions are not given independently of which roles they play. Instead, these metaphysical roles further constrain what spacetime points and regions can be. That is, spacetime points and regions are exclusively characterized by what they do. It is certainly unusual to assign spacetime regions the metaphysical role of instantiating ordinary properties, but it is not revisionist. By adding this metaphysical role to a region's repertoire, the standard conception of a region is not replaced by a different one; the standard conception is rather enriched. Spacetime regions are much thicker than we previously thought they were.

Furthermore, given that spacetime is an ontological black box, we cannot hope for much illumination of what it is for a spacetime region to be occupied by an ordinary object, and of what it is for a spacetime region

[2] Parsons (1990: ch. 10) discusses the idea that 'Suzie is clever' implicitly quantifies over states similarly to how 'Suzie walks' implicitly quantifies over events. In this picture, states receive a serious semantic role. While Parson's semantic account of temporal predication in terms of states differs from my representational account, the construal of temporal states and their subjects proposed above may be viewed as the metaphysical basis of both accounts of temporal predication.

to have an ordinary property simpliciter. As long as spacetime regions are primitives, these features of spacetime regions best remain primitives as well. Nonetheless, a gloss on regional instantiation in terms of temporal states is available that provides an intuitive way of thinking about this account of spatiotemporal instantiation that does not rely on the notion of a spacetime region. As sketched earlier, we may think of a spacetime region that is F simpliciter as a temporal state of being F. Moreover, we may think of an object that occupies a region that is F simpliciter as being in such a temporal state. Phrased in state-talk, the temporal-regions account of temporal supervenience says that the fact that *a* is F at different times is grounded in the fact that *a* is in different temporal states of being F.

As a corollary to the claim that spacetime regions are much thicker than we previously thought they were comes the claim that objects occupying spacetime regions are much thinner than we previously thought they were. At the spacetime level, the doctrine of regional instantiation strips ordinary objects of all their intrinsic properties; all properties by which we ordinarily characterize an object intrinsically at some time or other end up being properties of spacetime regions occupied by the object, without being properties of the object itself. Those who prefer ascribing ordinary properties to the contents of spacetime, as opposed to spacetime itself, might grant that shape is special, in that objects inherit their shapes from the regions they occupy, assuming that regions have shapes simpliciter. According to the temporal-regions account, however, shape is not special; an object inherits its entire intrinsic nature from its spatiotemporal location.

As regards the intuitive status of this claim, it must be emphasized that an object's having no mass or colour simpliciter is no more counter-intuitive than a spacetime region's having these properties simpliciter. Recall the earlier point that our intuitions are confined to ordinary time, and hence temporally modified. While no occupant of spacetime has a mass or colour simpliciter, according to the temporal-regions account, occupants of spacetime certainly have a mass and a colour at all times of their existence. Hence, to say that no occupant of spacetime has any ordinary intrinsic properties simpliciter does not go against what we ordinarily think.

If, at the level of spacetime, ordinary objects are intrinsically indistinguishable, how are they individuated? We distinguish intrinsically bare objects in spacetime by their path through spacetime: objects *x* and *y* are identical iff *x* and *y* occupy the same spacetime regions. Those who find intrinsically indistinguishable entities that are individuated by their position in the spacetime structure suspect should remember that, in the substantivalist tradition, spacetime points are intrinsically indistinguishable entities that are individuated by their position in the spacetime structure. Thus, if spacetime points in the familiar substantivalist world are not metaphysically suspect, then the occupants of spacetime in the three-dimensionalist world are not metaphysically suspect either.

The three-dimensionalist recommends a shift of metaphysical roles. Certain roles that used to be played by the occupants of spacetime are now played by the constituents of spacetime. Does this shift make the occupants of spacetime redundant? Think again of an instantaneous region that is F simpliciter as a temporal state of being F. The three-dimensionalist says that something is a subject of a temporal state of being F in virtue of occupying this state. Since spacetime itself, while providing the temporal states, does not provide the subjects of these states, objects occupying spacetime regions are not redundant in the three-dimensionalist world. In the three-dimensionalist's spacetime conception, an object plays the role of being the subject of temporal states. Without objects there would only be subject-less states. But the idea of a state, say a state of happiness, without a subject is just as incoherent as the idea of a football match without participants.

Where are we now? Our first complete picture of temporal supervenience was the four-dimensionalist temporal-parts account developed in Chapter 4. We are now presented with an alternative picture of temporal supervenience: the three-dimensionalist temporal-regions account. As an explanation of temporal supervenience, the temporal-regions account is at least as successful as the temporal-parts account. Both accounts presuppose the representational account of temporal predication that was shown to be superior to the relational account and the intensional account discussed in Chapter 3. I believe that the temporal-regions account and the temporal-parts account provide the most promising explanations of temporal supervenience; they are in the final round. Since they are rivals—one is based on three-dimensionalism, the other on four-dimensionalism—a decision must be made between them. In the previous chapter I made a case against the temporal-parts account. In the remainder of this chapter I shall make a case in favour of the temporal-regions account. This involves two tasks: first, to show that the temporal-regions account avoids the difficulties that threaten the temporal-parts account, and, secondly, to defend the temporal-regions account against objections. The first of these tasks will be taken up in the following section.

5.3 TEMPORAL REGIONS VERSUS TEMPORAL PARTS

Let us compare the temporal-regions account to the temporal-parts account. The two accounts are similar in an obvious respect: both accounts make use of the representational account of temporal predication, and accordingly both accounts construe the metaphysical basis of having a property at a time in terms of instantaneous representatives of ordinary objects that have properties simpliciter. Owing to this feature, the three-dimensionalist is able to offer an account of the metaphysical nature of change with the same virtues as the four-dimensionalist's account. Lewis famously argued that, in order to explain the possibility of change without temporal parts, one must either turn apparent properties, such as shapes, into relations of objects and times, or deny the reality of the past and the future.[3] The temporal-regions account shows that this assessment is incorrect. The account of change in terms of temporal regions, just as the account in terms of temporal parts, allows shapes to be genuine properties by letting shapes be instantiated simpliciter by instantaneous entities. Moreover, the account in terms of temporal regions, just as the account in terms of temporal parts, presupposes eternalism, the view that past, present, and future are equally real. Ordinary objects occupy multiple regions in a system of ordered hyperplanes, which forms the supervenience base of eternalist ordinary time. Hence, as regards the task of explaining change, the temporal-regions account and the temporal-parts account are equally powerful.

The two accounts differ in two respects: (*a*) the temporal representatives of ordinary objects are different kinds of things: in one case they are spacetime regions, in the other case they are occupants of spacetime regions; and (*b*) an ordinary object stands in different relations to its representatives: in one case an ordinary object occupies its representatives, in the other case an ordinary object has its representatives as parts. The temporal-parts account was stated in terms of moderate four-dimensionalism, according to which ordinary objects are distinct from and occupy spacetime regions. There is, however, an alternative version of four-dimensionalism, extreme four-dimensionalism, according to which ordinary objects are just extended spacetime regions. As we saw in Section 4.4, what speaks in favour of extreme four-dimensionalism is ontological parsimony. Given that ordinary objects are the mirror image of temporally extended spacetime regions, the roles played by objects and their parts might as well be played by spacetime regions and their parts. If moderate four-dimensionalism is replaced by extreme four-dimensionalism, then the moderate four-dimensionalist's account of spatiotemporal instantiation is naturally replaced by regional instantiation: instantaneous spacetime regions, temporal parts of ordinary objects, have ordinary properties simpliciter. As a consequence, difference (*a*) between the temporal-regions account and the temporal-parts account is not essential: there is a version of the temporal-parts account—the version

[3] See Lewis (1986*a*: 202–4; 1988). See Sect. 1.4 for further references and statements of the problems of change.

that employs extreme four-dimensionalism—according to which the temporal representatives of ordinary objects are spacetime regions, just as the temporal representatives on the temporal-regions account are spacetime regions. The divide between the two accounts thereby becomes smaller. This convergence is relevant in the dialectical arena in which the temporal-regions account is being discussed. When evaluating criticism of regional instantiation one must be aware that the temporal-parts account is the main competitor of the temporal-regions account, and that the temporal-parts account is driven towards the very same account of spatiotemporal instantiation.

The divide between the two accounts does not become so small, however, as to commit the temporal-regions account to a radical form of substantivalism. To think otherwise is to conflate having a property simpliciter and having a property at a time—that is, to conflate spacetime facts with ordinary temporal facts. The radical form of substantivalism to which the extreme four-dimensionalist commits herself says that ordinary objects are spacetime regions. Ordinary objects are the referents of everyday singular terms. They are the things that we describe in temporally modified language, the things that are persons or tables at a time, the things that are happy or heavy at a time. According to the temporal-regions account, these ordinary objects are not spacetime regions. They are rather, as (3D) states explicitly, things that occupy spacetime regions, and hence are distinct from such regions. The three-dimensionalist's material world is therefore something over and above substantivalist spacetime. It is 'contained' in spacetime, as traditional substantivalism has it.

By refraining from identifying ordinary objects with spacetime regions, the three-dimensionalist avoids the prima facie charge against extreme four-dimensionalism to be at odds with common sense: since persons are spacetime regions, spacetime regions are the way persons are and do the things persons do. For example, spacetime regions can be happy at a time. But surely spacetime regions are not the kinds of things that can meaningfully be said to be happy at any time? It is important to distinguish the question of the intuitive status of extreme four-dimensionalism from the question, addressed in the previous section, of the intuitive status of regional instantiation that extreme four-dimensionalism shares with three-dimensionalism. The relevant thing to keep in mind is that, while the present objection concerns spacetime regions appearing as subjects of temporal predications, according to regional instantiation spacetime regions are merely subjects of atemporal predications. Spacetime regions are only persons or tables simpliciter, and they are only happy or heavy simpliciter. Since our intuitions are confined to the level of ordinary time, regional instantiation avoids conflict with ordinary thought and talk.

While extreme four-dimensionalism's apparent clash with common sense can be explained away, extreme four-dimensionalism's failure to provide a plausible explanation of spatial supervenience constitutes its downfall, and hence commits the four-dimensionalist to the moderate version of her view (see Section 4.4). But moderate four-dimensionalism has problems of its own. In Chapter 4 I pointed to three such problems. One of them is that moderate four-dimensionalism postulates a redundant ontological category, since from the perspective of spacetime the work done by ordinary objects can equally be done by spacetime regions. Three-dimensionalism is better off than moderate four-dimensionalism in this respect, since in a three-dimensionalist spacetime the work done by ordinary objects cannot be done by spacetime regions alone.

Another problem for moderate four-dimensionalism is its failure to capture certain asymmetries between ordinary space and time. At the level of spacetime, the four-dimensionalist finds temporal as well as spatial extension, and temporal as well as spatial parts. The problem is that this space–time symmetry translates to the level of ordinary space and time, which clashes with our common conception of how objects are in space and time. The three-dimensionalist, by contrast, is able to save the ordinary space–time asymmetry by giving ordinary spatial and temporal facts a spatiotemporal supervenience base that is itself characterized by an asymmetry between its spatial and temporal aspects. The advantage of three-dimensionalism concerning ontological parsimony and space–time asymmetry was already apparent in Chapter 4, because it did not require the construction of a full three-dimensionalist account of temporal supervenience. The case of the third problem is different.

The problem that lies at the heart of the temporal-parts account of temporal supervenience is the problem of predicational overkill. The temporal-parts account says that *a* is F at *t* iff *a* has a temporal part that occupies a member-region of *t* and that is F. This account has the consequence that temporal predication is closed under the temporal-part relation in the following sense. Since all that is required to be F at *t* is to have a temporal part, a_t, that occupies a member-region of *t* and that is F, every object that has a_t as a temporal part is F at *t*. Hence, if *a* is F at *t*, then every object that overlaps with *a* in that it has a_t as a temporal part is also F at *t*. In Section 4.5, closure under parthood was shown to have strongly counter-intuitive consequences arising from the theory of temporal parts. The difficulty, in a nutshell, is that the theory of temporal parts entails that, if an ordinary object *a* has a temporal part that is F simpliciter, then there are many things distinct from *a* that also have that temporal part. By closure under parthood, it follows that a persisting object cannot be the only thing that has a given property at a time; many other things will also have the property at that time if the persisting object does. An analogous problem was shown to arise for the temporal-counterparts account of temporal supervenience and its consequence that temporal predication is closed under the temporal-counterpart relation.

The temporal-regions account of temporal supervenience shares none of these defects with the temporal-parts account and the temporal-counterparts account. The reason is simple. The essential difference between the temporal-regions account and the temporal-parts/temporal-counterparts account is that, according to the latter, all ordinary facts of temporal instantiation logically supervene on mereological facts about an object's temporal parts/temporal counterparts, whereas, according to the temporal-regions account, most facts of temporal instantiation logically supervene on non-mereological facts about an object's spatiotemporal location. Accordingly, the temporal-regions account is closed neither under parthood nor under counterparthood. It is rather *closed under spatiotemporal occupation*: every object that occupies an instantaneous spacetime region that is F simpliciter and that is a member of *t* is F at *t*. This neutrality from mereological considerations is what prevents the proliferation of unwanted facts of temporal instantiation in the case of the temporal-regions account. If *a* has a temporal part that is F simpliciter, it immediately follows, in the four-dimensionalist picture, that many other things, proper parts of *a*, also have that temporal part. But, if *a* occupies a region that is F simpliciter, it does not immediately follow, in the three-dimensionalist picture, that many other things also occupy that region. Hence, the temporal-regions account is not counter-intuitive in the way the temporal-parts account and the temporal-counterparts account are. To be sure, closure under occupation raises a number of issues on its own, which will be discussed in Section 5.6. But these issues are another story. What matters right now is that the temporal-regions account avoids predicational overkill, which constitutes the central advantage of this three-dimensionalist account of temporal supervenience.

Having given three arguments in favour of the temporal-regions account—the argument from predicational overkill, the argument from space–time asymmetry, and the argument from ontological parsimony—I will, in the remainder of this chapter, discuss three apparent problems: the problem of special relativity, the problem of predicate abstraction, and the problem of coincidence.

5.4 THREE-DIMENSIONALISM AND SPECIAL RELATIVITY

The problem of relativistic supervenience, as stated in Section 1.6, is to determine which spacetime facts the facts of relativistic time logically supervene on, and to explain how they supervene. In this section I will show how the three-dimensionalist temporal-regions account as well as the four-dimensionalist temporal-parts account of temporal supervenience can be reconfigurated to serve as accounts of relativistic supervenience.

The task of explaining relativistic supervenience is particularly pressing for the three-dimensionalist in the light of the widespread view that Special Relativity (SR) favours the four-dimensionalist outlook. One motivation for this claim is the combination of the thesis that SR requires eternalism and the thesis that eternalism requires four-dimensionalism. Although a strong case can be made in favour of the first thesis, the second thesis is false, as has been shown in Section 2.4. So we have already at least partly demotivated the claim that SR works against three-dimensionalism.[4] But to demotivate this claim is not to show that it is false. And to show that the claim is false is to demonstrate the compatibility of the three-dimensionalist picture with SR. This is the main task of this section.

Relativistic supervenience concerns the relationship between relativistic time and Minkowski spacetime. Minkowski spacetime is the metaphysical and explanatory basis of relativistic time. All facts about relativistic time logically supervene on facts about Minkowski spacetime. This is the thesis of relativistic supervenience. Among the facts about relativistic time are facts of temporal persistence and variation and of relativistic persistence and variation. The relativistic analogues of ordinary temporal persistence and variation are facts concerning an object's existence and its instantiation of properties at various times relative to the same frame of reference: *a* is F at t_1 relative to f and *a* is not F at t_2 relative to f, where t_1 is earlier than or later than t_2 relative to f. The genuinely relativistic facts of relativistic persistence and variation concern an object's existence and its instantiation of properties at various times across different frames of reference: *a* is F at t_1 relative to f_1 and *a* is not F at t_2 relative to f_2, where t_1 and t_2 are individuated relative to different frames and bear no frame-relative temporal relations to each other. The problem of relativistic supervenience is to determine the spatiotemporal supervenience base of these facts of temporal persistence and variation and of relativistic persistence and variation in the context of SR, and to explain how they supervene.

[4] For more on this sort of demotivation, see Rea (1998).

The first step is to give (3D) and (4D) relativistically acceptable reformulations. (3D) and (4D) themselves are relativistically unacceptable, since they are stated in terms of the notions of a temporally unextended, or instantaneous, and a temporally extended spacetime region, which presuppose absolute simultaneity between spacetime points (see (IR) and (ER)). Relativistic versions of the notions of a temporally unextended spacetime region and a temporally extended spacetime region were stated as (IR_{SR}) and (ER_{SR}) in Section 2.1. Presupposing these definitions, (3D) and (4D) were given relativistically acceptable updates:

($3D_{SR}$) (i) an ordinary object occupies multiple spacetime regions, and (ii) these spacetime regions are temporally unextended, or instantaneous, relative to some inertial frame of reference, and lie on different frame-relative hyperplanes.

($4D_{SR}$) (i) an ordinary object occupies a unique spacetime region, and (ii) this spacetime region is composed of regions that are temporally extended relative to some inertial frame of reference.

Just as (4D) can be developed into a theory of temporal parts of objects, so ($4D_{SR}$) can be developed into a theory of frame-relative temporal parts of objects. In order to state a relativistic theory of temporal parts, we may first define the notion of a frame-relative temporal part of a spacetime region, and then define the notion of a frame-relative temporal part of an object in terms of the latter:

($TP{:}R_{SR}$) A spacetime point or region R is a *temporal part* of a spacetime region R' relative to a frame of reference $f =_{df}$ R is a maximal sum of parts of R' that are simultaneous relative to f.[5]

($TP{:}O_{SR}$) An object x is a *temporal part* of an object y relative to a frame of reference $f =_{df}$ (i) x is a part of y, (ii) y occupies a spacetime region R, (iii) x occupies a point or region that is a temporal part of R relative to f, and (iv) x does not occupy any other region.

To these definitions, the four-dimensionalist may add the *doctrine of arbitrary frame-relative temporal parts*:

(ATP_{SR}) If an ordinary object x occupies a spacetime region R, then for every frame-relatively unextended temporal part R' of R there is a part x' of x such that (i) x' occupies R', and (ii) x' does not occupy any other spacetime region.

[5] This definition may be spelled out as follows: a spacetime point or region R is a *temporal part* of a spacetime region R' relative to $f =_{df}$ (i) R is a part of R', (ii) the spacetime points in R are all simultaneous relative to f, and (iii) no spacetime point that is in R' but not in R is simultaneous with all spacetime points in R relative to f.

Given (TP:O_{SR}), it follows from ($4D_{SR}$) and (ATP_{SR}) that ordinary objects have proper frame-relative temporal parts. So much for a brief sketch of a relativistic theory of temporal parts.

Given ($3D_{SR}$) and ($4D_{SR}$) as different relativistic accounts of an ordinary object's spatiotemporal location, relativistic three-dimensionalist and four-dimensionalist accounts of spatiotemporal instantiation, of how ordinary properties are distributed across Minkowski spacetime, are not far away. The three-dimensionalist may transpose her temporal-regions account of spatiotemporal instantiation by saying that the things that have ordinary properties simpliciter are frame-relatively instantaneous spacetime regions occupied by ordinary objects. The four-dimensionalist may likewise transpose her temporal-parts account of spatiotemporal instantiation by saying that the things that have ordinary properties simpliciter are frame-relative temporal parts of ordinary objects.

Assuming these relativistic accounts of spatiotemporal location and instantiation, we may specify which facts about Minkowski spacetime the facts of temporal and relativistic persistence and of temporal and relativistic variation logically supervene on. Let us start with the three-dimensionalist's account. According to the three-dimensionalist, an object's temporal and relativistic persistence is entailed by the object's occupation of certain frame-relatively instantaneous regions. Moreover, an object's temporal and relativistic variation in properties is entailed by the object's occupation of certain frame-relatively instantaneous regions that have incompatible properties simpliciter. So far, so good. But temporal persistence clearly differs from relativistic persistence. Likewise for temporal and relativistic variation. How is this difference reflected at the spacetime level?

Contrast a pole *P*'s temporal variation in temperature with the same pole's relativistic variation in length.[6] The fact that *P* has temperature T_1 at t_1 relative to f and temperature T_2 at t_2 relative to f is entailed by the fact that *P* occupies a region R_1 that has T_1 and another region R_2 that has T_2, where R_1 and R_2 lie on different hyperplanes of simultaneity relative to the *same* frame of reference f. On the other hand, the fact that *P* has length L_1 at t_1 relative to f and length L_2 at t_3 relative to f' is entailed by the fact that *P* occupies a region R_1 that has L_1 and another region R_3 that has L_2, where R_1 and R_3 lie on different hyperplanes of simultaneity relative to *different* frames of reference f and f'. This difference concerning the relationship of R_1 and R_2 in the case of temporal variation and the relationship of R_1 and R_3 in the case of relativistic variation may be illustrated by Figure 5, in which (x, t) corresponds to f and (x', t') corresponds to f':

[6] For a brief characterization of the phenomenon of length contraction, see Sect. 1.6.

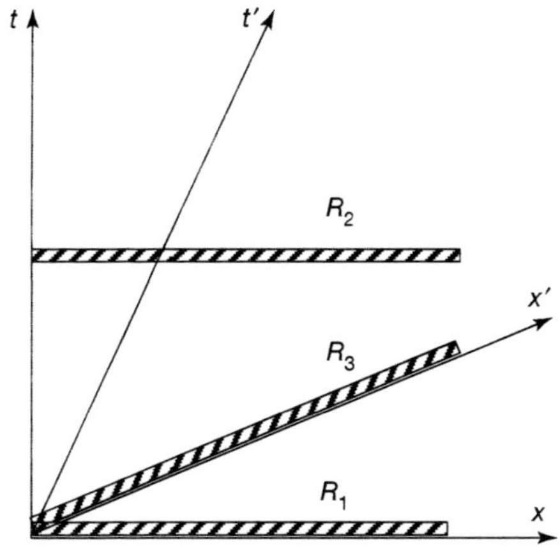

Figure 5

What remains to be given is an explanation of the link between occupying a region that is F simpliciter and that lies on some hyperplane relative to some frame, and being F at a time relative to a frame. Consider the relativistic temporal predication 'a is F at t_i relative to f'. First, this predication may be simplified to 'a is F at t_{if}'. Next, t_{if} may be construed as the maximal set of spacetime points and regions that lie on hyperplane i relative to frame f. The third and final step in the explanation is the following relativistic counterpart of (TS_{3D}):

(RS_{3D}) Necessarily, a is F at t_{if} iff a occupies a spacetime region that is F simpliciter and that is a member of t_{if}.

This completes the three-dimensionalist temporal-regions account of relativistic supervenience.

The four-dimensionalist takes analogous steps in her explanation of relativistic supervenience. Given frame-relative simultaneity, the four-dimensionalist may specify the following spacetime facts as underlying a pole P's temporal variation in temperature and relativistic variation in length. The fact that P has temperature T_1 at t_1 relative to f is entailed by the fact that P has a temporal part relative to f that occupies region R_1 and has T_1, and the fact that P has temperature T_2 at t_2 relative to f is entailed by the fact that P has another temporal part relative to f that occupies R_2 and has T_2, where R_1 and R_2 lie on different hyperplanes of simultaneity relative to the *same* frame of reference f. On the other hand, the fact that P has length L_1 at t_1 relative to f is entailed by the fact that P has a temporal part relative to f that occupies R_1 and has L_1, and the fact that P has length

L_2 at t_3 relative to f' is entailed by the fact that P has another temporal part relative to f' that occupies R_3 and has L_2, where R_1 and R_3 lie on different hyperplanes of simultaneity relative to *different* frames of reference f and f'. (For illustration of this difference concerning the relationship of R_1 and R_2 in the case of temporal variation and the relationship of R_1 and R_3 in the case of relativistic variation, see again Figure 5.)

In order to arrive at a full account of relativistic supervenience, the four-dimensionalist is required to provide an explanatory bridge from having a temporal part that is F simpliciter and that occupies a region that lies on some hyperplane relative to some frame, to being F at a time relative to a frame. The first two steps in giving such an explanation are the same as in the three-dimensionalist case: '*a* is F at t_i relative to f' is read as '*a* is F at t_{if}', and t_{if} is the maximal set of spacetime points and regions that lie on hyperplane *i* relative to frame f. The third step in the explanation is the following relativistic counterpart of (TS_{4D}):

(RS_{4D}) Necessarily, *a* is F at t_{if} iff *a* has a temporal part relative to f that is F simpliciter and that occupies a spacetime region that is a member of t_{if}.

This completes the four-dimensionalist temporal-parts account of relativistic supervenience.

The foregoing discussion of relativistic supervenience should leave no doubt that the three-dimensionalist picture is perfectly compatible with SR. Even granting this compatibility, one might still view SR as favouring four-dimensionalism over three-dimensionalism. Yuri Balashov considers relativistic variation in shape and asks what underlies such variation at the level of spacetime—or, as I would put it, he asks for the spatiotemporal supervenience base of relativistic variation in shape. Balashov points out that, according to the four-dimensionalist, different 3D shapes that an object has at different times relative to different frames are, at the level of spacetime, cross-sections of a single four-dimensional object, that '[lend] themselves to an arrangement in a compact and smooth 4D volume' (1999: 653). Balashov compares this situation to

> the way in which usual three-dimensional objects in space stand behind all their perspectival plane projections. A set of pictures of a house, say, taken from different vantage points display a considerable diversity of two-dimensional shapes. But behind this diversity stands the invariant three-dimensional shape of the house itself. Similarly, the pole displays a variety of lengths in different space–time perspectives. But behind this variety stands the invariant four-dimensional shape of the perduring whole. (Balashov 2000: 334)

He then goes on to claim that the three-dimensionalist has a hard time explaining how 'separate and loose' 3D shapes come together in such a remarkable and harmonious unity: 'where the four-dimensionalist has a ready and natural explanation of this fact [...] the three-dimensionalist must regard it as a brute fact, indeed, as a complete mystery' (1999: 653). In order to evaluate Balashov's point, we must ask what exactly the explanatory task is that the three-dimensionalist allegedly fails to meet. Balashov requires the three-dimensionalist to explain how 3D shapes are spatiotemporally 'unified', in the sense that different 3D shapes correspond to different perspectives on a 4D object. The problem with this requirement is that it is entirely unjustified. The main thing to be explained about relativistic variation is its spatiotemporal nature. The three-dimensionalist temporal-regions account manages to explain just that. The alleged problem for the three-dimensionalist appears when a constraint is introduced on what form the supervenience base of relativistic variation should take, the constraint being that the 3D shapes form a spatiotemporal 'unity', in the sense, to repeat, that different 3D shapes correspond to different perspectives on a 4D object. Unlike relativistic variation, however, this unity has no pre-theoretic significance. Balashov explicitly takes the spatial case of 2D shapes of a 3D object as his guide to relativistic variation. He straightforwardly extrapolates from the spatial to the relativistic temporal case. But there is no pre-theoretic reason to believe that the relativistic temporal case should behave at all like the spatial case. Instead of being a general phenomenon to be explained by various theories, the spatiotemporal unity of 3D shapes is rather just a feature of a particular theory, namely four-dimensionalism. As a consequence, the point that the three-dimensionalist cannot explain this unity begs the question. It amounts to saying that four-dimensionalism cannot be explained in terms of three-dimensionalism.[7] I conclude that SR, in particular the phenomenon of relativistic variation, poses no threat to three-dimensionalism.

5.5 TEMPORAL PREDICATION WITH COMPLEX PREDICATES

A further apparent problem for the temporal-regions account consists in the incompatibility of the latter with the classical principle of predicate abstraction. The principle of predicate abstraction says that we can form

[7] It is also worth pointing out that Balashov works only with three-dimensionalist accounts of relativistic variation that do not exhaust the space of options, and that do not work for stronger reasons than he gives. For these accounts are relativistic variants of the relational and intensional accounts of temporal predication considered and rejected in Chapter 3.

a complex predicate from any open sentence. Any sentence containing a singular term has an equivalent simple predication with this term in subject position. For instance, 'Charlie is losing his mind', is equivalent to 'Charlie is such that he is losing his mind' which contains the complex predicate 'is such that he is losing his mind'. In general, for any sentence '$S(a)$',

(PA) $\Box[S(a) \equiv \lambda x[S(x)](a)]$

Consider the identity statement 'Charlie is identical to Charlie'. According to (PA), this statement is equivalent to 'Charlie is such that he is identical to Charlie'. Letting 'c' stand for Charlie, we have the following instance of (PA):

$$(1)\ \Box[c = c \equiv \lambda x[x = c](c)]$$

The left-hand side of (1) contains the simple two-place predicate of identity. The right-hand side of (1) contains the complex one-place predicate '$\lambda x[x = c](\ )$', which is obtained from the two-place predicate '$=$' (or '$\lambda x\lambda y[x = y](\ ,\ )$') by filling up one of its places by the constant 'c'. The intuitive difference between the left-hand side and the right-hand side of (1) may be captured in the language of properties and relations: the left-hand side says that Charlie has the relation of identity to Charlie, whereas the right-hand side says that Charlie has the property of being identical to Charlie. The property of being identical to Charlie is often classified as a relational property in the sense that Charlie's having it somehow consists in having the relation of identity to himself. In the present context, such talk of properties and relations serves merely a heuristic function. The argument to be given does not require taking a stand on whether there is a relational property corresponding to the predicate '$\lambda x[x = c](\ )$', or, more generally, whether there is a complex property corresponding to any of the complex predicates the predicate-abstraction principle allows us to form. (I shall return to the issue of complex properties at the end of this section.)

Now to temporal predication. The predicate-abstraction principle (PA) has a temporally modified variant, according to which 'At t, Charlie is losing his mind' is equivalent to 'At t, Charlie is such that he is losing his mind'. The temporal-regions account incorporates the assumption that temporal operators function as sentence modifiers. Given this assumption, the temporally modified variant of (PA) may be stated as follows:

(PA_t) $\Box[\text{At } t[S(a)] \equiv \text{At } t[\lambda x[S(x)](a)]]$

Consider the temporally modified identity statement 'At t, Charlie is identical to Charlie'. According to (PA_t), this statement is equivalent to 'At t, Charlie is such that he is identical to Charlie'. Or formally:

$$(2)\ \Box[\text{At } t[c = c] \equiv \text{At } t[\lambda x[x = c](c)]]$$

This equivalence creates trouble for the temporal-regions account. To see why, let us remind ourselves of the account's core principle:

(TS_{3D}) $\Box[\text{At } t[\text{F}(a)] \equiv \exists R(\text{IR}(R) \,\&\, \text{O}(a, R) \,\&\, R \in t \,\&\, \text{F}(R)\,)]$

Let us apply (TS_{3D}) to the left-hand side and to the right-hand side of (2). The left-hand side contains a two-place predicate and therefore is an instance of the dyadic version of (TS_{3D}):

(3) $\Box[\text{At } t[c = c] \equiv \exists R(\text{IR}(R) \,\&\, \text{O}(c, R) \,\&\, R \in t \,\&\, R = R]$

The right-hand side of (2) contains a one-place predicate and therefore is an instance of the monadic (TS_{3D}) itself:

(4) $\Box[\text{At } t[\lambda x[x = c](c)] \equiv \exists R(\text{IR}(R) \,\&\, \text{O}(c, R) \,\&\, R \in t \,\&\, \lambda x[x = c](R)]$

Since it is clearly possible for Charlie to occupy a member-region of t that is identical to itself, (3) is unproblematic. (4) has a different status, though. Since Charlie is distinct from the spacetime region he occupies, it is inconsistent to say that R has the property of being identical to Charlie—$\lambda x[x = c](R)$—because only Charlie has the property of being identical to Charlie. (I am assuming that '=' is to be read as strict identity, as opposed to some weaker relation.) This makes 'At $t[\lambda x[x = c](c)]$' trivially false; and so (2) is false. Hence, (TS_{3D}) is incompatible with predicate abstraction.

The threat of predicate abstraction is not confined to the case of identity. Suppose that, at t, Suzie is bored of Charlie. According to (PA_t), this statement is equivalent to 'At t, Suzie is such that she is bored of Charlie'. Abbreviating the two-place predicate 'is bored of' as 'B(,)', we have the following instance of (PA_t):

(5) $\Box[\text{At } t[\text{B}(s, c)] \equiv \text{At } t[\lambda x[\text{B}(x, c)](s)]\,]$

By the dyadic version of (TS_{3D}),

(6) $\Box[\text{At } t[\text{B}(s, c)] \equiv \exists R \exists R'(\text{IR}(R) \,\&\, \text{IR}(R') \,\&\, \text{O}(s, R) \,\&\, \text{O}(c, R') \,\&\, R \in t \,\&\, R' \in t \,\&\, \text{B}(R, R')\,)]$

By (TS_{3D}),

(7) $\Box[\text{At } t[\lambda x[\text{B}(x, c)](s)] \equiv \exists R(\text{IR}(R) \,\&\, \text{O}(s, R) \,\&\, R \in t \,\&\, \lambda x[\text{B}(x, c)](R)]$

Since it is consistent to suppose that it is true that R has the relation of being bored-of to R'—$\text{B}(R, R')$—but false that R has the property of being bored of Charlie—$\lambda x[\text{B}(x, c)](R)$—(5) is false.

The case of identity and the case of boredom are both cases in which a seemingly unproblematic ordinary temporal predication containing a complex predicate obtained from (PA_t) is, by (TS_{3D}), equivalent to a problematic atemporal predication—see (4) and (7). A further case runs in the opposite direction: a seemingly unproblematic atemporal predication containing a complex predicate obtained from (PA) is, by (TS_{3D}), equivalent to a problematic temporal predication. The following case is of this sort. Suppose that Zoe is happy at *t*. By the temporal-regions account, this entails that Zoe occupies a region *R* that is a member of *t* and that is happy simpliciter. Since *R* is also a spacetime region simpliciter, it follows that Zoe is a spacetime region at *t*, which seems counter-intuitive. The main thing to notice here is that the predicate 'is a spacetime region' is a complex one-place predicate derived by (PA) from a two-place predicate: to say that *R* is a spacetime region is really just short for saying that *R* is an extended part of spacetime.[8] The two-place predicate avoids the trouble caused by the complex one-place predicate. Consider the dyadic version of (TS_{3D}): necessarily, *a* is G to *b* at *t* iff *a* occupies a region *R* and *b* occupies a region *R'*, where both *R* and *R'* are members of *t*, and *R* is G to *R'*. Accordingly, for Zoe to be a part of spacetime at *t*, spacetime itself would have to occupy an instantaneous region. Since this is impossible, Zoe cannot be a part of spacetime at a time. The point is that the intuitively problematic result that Zoe is a spacetime region at a time obtains only if 'is a spacetime region' is taken as a complex one-place predicate, and so the result rests on predicate abstraction. A similar problem arises for the predicate 'is instantaneous'. Assuming that Zoe occupies a region that has the property of being instantaneous and that lies on *t*, it follows by (TS_{3D}) that Zoe is instantaneous at *t*, which seems counter-intuitive. The predicate 'is instantaneous' is a complex predicate: to say that a region is instantaneous is short for saying that the region is made up of simultaneous points. As in the case of 'is a spacetime region', the problematic result rests on predicate abstraction. Zoe ends up being instantaneous at a time only if 'is instantaneous' is construed as a complex one-place predicate.[9]

[8] Remember that we started by saying that there is this thing, spacetime. This thing has parts, primitive unextended parts and extended parts made up of the unextended parts. Then we introduced the terms 'spacetime point' and 'spacetime region' for the unextended parts and the extended parts of spacetime, respectively.

[9] One might deny that Zoe's being a spacetime region at a time is counter-intuitive, on the ground that we have no reliable intuitions about spacetime regions. Our lack of such intuitions, one might claim, is due to the fact that the concept of a spacetime region is a technical concept, a concept that plays no role in ordinary thought and talk. I mentioned this sort of reply in Sect. 4.4 when discussing the intuitive status of extreme four-dimensionalism.

The threat of predicate abstraction is not confined to the temporal-regions account. It applies equally to the temporal-parts account, and therefore has no effect on the debate between the two accounts. In fact, the problem arises, in one variant or the other, for any account of *a*'s having a property at a time that ascribes the property to an entity that is distinct from *a*. In order to get a sense of the generality of the problem, it will suffice quickly to go through the identity case again, this time working with (TS_{4D}). Applying (TS_{4D}) to the left-hand side and the right-hand side of (2) yields the following equivalences (I shall state the latter semi-formally this time):

(8) $\Box$[At $t[c = c] \equiv c$ has a temporal part, c_t, that is located at t & $c_t = c_t$]

(9) $\Box$[At $t[\lambda x[x = c](c)] \equiv c$ has a temporal part, c_t, that is located at t & $\lambda x[x = c](c_t)$]

Since it is clearly possible for Charlie to have a temporal part located at t that is identical to itself, (8) is unproblematic. (9), on the other hand, is problematic. Since Charlie is a temporally extended person, Charlie's instantaneous temporal part c_t is a proper part of Charlie and hence distinct from him. It is therefore inconsistent to say that c_t has the property of being identical to Charlie—$\lambda x[x = c](c_t)$—because only Charlie has the property of being identical to Charlie. This makes 'At $t[\lambda x[x = c](c)]$' trivially false; and so (2) is false. Hence, (TS_{4D}) as well as (TS_{3D}) is incompatible with predicate abstraction.

A radical response to this result by the friends of temporal regions or of temporal parts would be to reject the principles of predicate abstraction, (PA) and (PA_t), as false. This is a bad idea, though, since the principle is supported by strong intuition when put in ordinary terms: to say that Suzie is bored of Charlie is surely equivalent to saying that Suzie is such that she is bored of Charlie—or, in property-speak, to say that Suzie bears the relation of boredom to Charlie is equivalent to saying that Suzie has the property of being bored of Charlie. Is there any other way out?

As an alternative response to the threat of predicate abstraction, one might reject the representational account of temporal predication. In Section 4.1, I discussed two alternatives to the representational account that still yield (TS_{4D}): the subject-modifier account and the ellipsis account. Both of these accounts might be thought to avoid the threat of predicate abstraction. The four-dimensionalist subject-modifier account construes 'at *t*' in '*a* is F at *t*' and in '*a* is G to *b* at *t*' as a subject modifier yielding '*a*-at-*t* is F' in the monadic case and '*a*-at-*t* is G to *b*-at-*t*' in the dyadic case. Given the theory of temporal parts, '*a*-at-*t*' may then be taken as an abbreviation of the complex definite description 'the temporal part of *a* that occupies a member-region of *t*'. On this proposal, the unregimented temporal predication 'At *t*, Charlie is identical to Charlie' becomes 'Charlie-at-*t* is identical to Charlie-at-*t*'. According to (PA), this is equivalent to 'Charlie-at-*t* is such that he is identical to Charlie-at-*t*':

$$(10)\quad \Box[c\text{-at-}t = c\text{-at-}t \equiv \lambda x[x = c\text{-at-}t](c\text{-at-}t)]$$

Since (10) is unproblematic for (TS_{4D}), the subject-modifier reading renders the temporal-parts account of temporal supervenience compatible with predicate abstraction. The four-dimensionalist ellipsis account of temporal predication is even simpler. It construes the left-hand side of (TS_{4D}) as elliptical for its right-hand side. Thus, 'At *t*, Charlie is identical to Charlie' is short for 'Charlie has a temporal part that occupies a member-region of *t*, c_t, such that c_t is identical to c_t'. Applying (PA) to '$c_t = c_t$' yields a variant of (10), and hence the ellipsis reading of temporal predications avoids the threat of predicate abstraction in the same straightforward way as the subject-modifier reading. Moreover, the subject-modifier account and the ellipsis account of temporal predication have obvious three-dimensionalist analogues, which means that these accounts also offer a reconciliation of (TS_{3D}) with predicate abstraction.[10]

This strategy of avoiding the threat of predicate abstraction is unsuccessful. The alleged way out consists in applying the principle of predicate abstraction only after the sentence 'At *t*, Charlie is identical to Charlie' has been regimented in subject-modifier style. The problem recurs, however, once we realize that the principle of predicate abstraction may be stated informally and applies to unregimented sentences as well as regimented ones. Before even thinking about the logic of temporal modifiers, it is clear that 'At *t*, Charlie is identical to Charlie' is equivalent to 'At *t*, Charlie is such that he is identical to Charlie'—or, in property-speak, 'At *t*, Charlie has the property of being identical to Charlie'. Applying the subject-modifier reading to this unregimented equivalence, yields the following result:

$$(11)\quad \Box[c\text{-at-}t = c\text{-at-}t \equiv \lambda x[x = c](c\text{-at-}t)]$$

Since this equivalence is just as troublesome as (2), the threat of predicate abstraction remains. Similarly for the ellipsis account.[11]

A move that really does avoid the threat of predicate abstraction is to reinterpret (PA) (and (PA_t)). The right-hand side of (PA) can be read more or less seriously. It may be taken as literal and, accordingly, as saying

[10] Assuming the subject-modifier reading, the three-dimensionalist interprets '*a*-at-*t*' as 'the region that is occupied by *a* and a member of *t*'. The three-dimensionalist ellipsis account simply says that the left-hand side of (TS_{3D}) is elliptical for its right-hand side.

[11] The appeal to the subject-modifier account and the ellipsis account is also questionable for reasons given in Sect. 4.1.

something that is different from the left-hand side. Or it may be taken as non-literal and, accordingly, as a misleading way of saying exactly the same thing as the left-hand side. If the right-hand side of (PA) is taken as non-literal, then the principle is trivially true. To see why the trivial (PA) is compatible with the temporal-regions account and the temporal-parts account, notice that the temporal predication that occurs on the left-hand side of (TS_{3D}) and (TS_{4D}) is required to be strictly and literally true. Now take the case of identity and reconsider (2). On both accounts, the right-hand side of (2) turns out to be trivially false. To apply the accounts to the right-hand side of (2) is to presuppose that the latter is strictly and literally true. But, if (PA) is trivial, then the right-hand side of (2) is non-literal, and so the accounts cannot be applied. The other cases discussed can be treated similarly. In each case, the problem caused by a particular predication containing a complex predicate is avoided on the grounds that (TS_{3D}) and (TS_{4D}) cannot be applied to such a predication, because such a predication is not to be taken as literal. These considerations suggest that the claim that predicate abstraction is incompatible with the temporal-regions account and the temporal-parts account rests on the mistake of taking a misleading statement at face value.

The suggestion to render (PA) (and (PA_t)) trivial may be given independent support by a case concerning modal predication. Consider the following puzzle about the relationship between (PA) and a simple principle concerning existence. Formalizing 'exist' as 'E( )', for any monadic predication 'F(a)',

(E) $\Box[F(a) \supset E(a)]$[12]

Analogously for n-adic predications. One might want to distinguish between different senses of 'exist'. I will focus on the sense that is definable from identity and the unrestricted existential quantifier: $E(a) =_{df} \exists x(a = x)$. Intuitively, that a exists means that a is something. Given this construal of the existence-predicate, the existence principle (E) is almost trivial. For how could a be F without being something that is F, and hence something? Here is how (E) clashes with (PA). Surely Socrates might not have existed. In possible-worlds talk this is to say that there is a possible world w at which Socrates does not exist: At $w[\neg E(\text{Socrates})]$. By (PA), this is equivalent to: At $w[\lambda x[\neg E(x)](\text{Socrates})]$.[13] By (E), it follows that At $w[E(\text{Socrates})]$, which

[12] Kit Fine calls this principle *Predicate Actualism*; see Fine (1985: 169). Replacing the left-hand side of (E), 'F(a)', with 'I(a, F-ness)', or informally, 'a instantiates F-ness', yields the principle that Alvin Plantinga calls *Serious Actualism*; see Plantinga (1983, 1985). I accept both principles but focus on (E).

[13] In the possible-worlds framework, (PA) becomes (PA_w): $\forall w[\text{At } w[S(a)] \equiv \text{At } w[\lambda x[S(x)](a)]]$.

contradicts our assumption. Some philosophers (for example, Fine and Pollock)[14] have suggested rejecting (E), others (for example, Plantinga)[15] rejecting (PA). Either way out is unsatisfactory, since both principles are supported by strong intuition. Fortunately, there is a solution that leaves each of them intact. The solution is to construe (PA) as trivial in the sense described earlier. Given this construal, 'At w, Socrates is such that he does not exist'—'At $w[\lambda x[\neg E(x)](\text{Socrates})]$'—turns out to be no more than a misleading paraphrase of 'At w, Socrates does not exist'—'At $w[\neg E(\text{Socrates})]$'. Since (E) requires its antecedent to be literal, it does not follow by (E) that Socrates exists at w. The conclusion that Socrates exists at w thus rests on the mistake of taking a misleading statement at face value.[16]

I shall conclude this section with a remark to the effect that the interpretation of (PA) is not just a semantic matter, but an issue with ontological significance. It is widely believed that, if there are properties, they somehow correspond to predicates. As a corollary, if there are complex properties, they correspond to complex predicates. Standard examples of complex predicates are conjunctive predicates such as '$\lambda x[F(x) \wedge G(x)](\,)$', disjunctive predicates such as '$\lambda x[F(x) \vee G(x)](\,)$', negative predicates such as '$\lambda x[\neg F(x)](\,)$', and relational predicates such as '$\lambda x[R(x, a)](\,)$'. Advocates of complex properties claim that a conjunctive predicate has a corresponding conjunctive property (such as being green and spherical), that a disjunctive predicate has a corresponding disjunctive property (such as being green or red), that a negative predicate has a corresponding negative property (such as non-existence), and that a relational predicate has a corresponding relational property (such as being identical to Zoe). It is obvious that for these correspondences to obtain it must be possible for a statement containing one of the mentioned complex predicates to be strictly and literally true. (PA) says that we can form a complex predicate from any open sentence in a way that is formally representable by λ-abstraction. As we have seen, to render (PA) trivial is to render any sentence containing a complex predicate formed in this way non-literal. All the standard complex predicates mentioned above have in common that they are formed by λ-abstraction. (Perhaps there are non-standard complex predicates that are not formed in this way. But these do not concern us here.) It therefore follows from the triviality of (PA) that the aforementioned

[14] See Fine (1985) and Pollock (1985).

[15] See Plantinga (1985).

[16] Fine makes a point that comes very close to this: 'One can so understand property-talk, that to say Socrates has the property of not existing is to say no more, in modal contexts, than that Socrates does not exist. It is then trivial, if it is possible that Socrates not exist, that it is possible that Socrates possess the property of not existing' (1985: 165). But Fine does not conclude that this reading of the predicate-abstraction principle is compatible with the existence principle. Instead, he rejects the existence principle.

complex predicates have no corresponding complex properties, and hence that, if properties correspond to predicates, then there are no conjunctive, disjunctive, negative, or relational properties.[17]

5.6 COINCIDENCE AND SORTAL RELATIVITY

Occupying a spacetime region simpliciter, or occupying a place at a time, plays a central role in the temporal-regions account of temporal supervenience, since an object inherits a property it has at a time from a spacetime region it occupies. Accordingly, cases in which distinct things occupy the same spacetime region, or occupy the same place at the same time—for short, cases in which distinct things *coincide*—are to be given particular attention when evaluating the temporal-regions account. In this section, I will discuss two kinds of apparent coincidence of distinct things. The first kind, which concerns universals and tropes and which I will go through rather quickly, does not present a problem for the temporal-regions account, but merely serves to mark a relevant metaphysical consequence of the latter. The second kind of coincidence, which concerns ordinary objects and which I will discuss in detail, does present a prima facie problem for the temporal-regions account.

Coincidence of universals and tropes

The first case of apparent distinct coincidents arises in connection with the question how material objects should be characterized in terms of ontological categories. Such a characterization can take two forms: either one can take objects to belong to a *sui generis* ontological category or one can take them to be metaphysical constructions from entities that belong to other ontological categories.[18] Since the view of material objects as metaphysical constructions is what brings up issues of coincidence, consider the following two well-known variants of this view. I shall confine myself to brief sketches.

Objects as one-category constructions (bundle theory)

The most economical plan is to take an object as constructed from entities belonging to a single *sui generis* ontological category, as some sort of bundle

[17] The existence of standard complex properties has been denied, though on different grounds, by Ramsey (1925) and Mellor (1991*b*, 1992).

[18] See Oliver (1996: 21).

of what will be called *tropes*.[19] Then we need a primitive bundling relation to say how the tropes that comprise the object are united. This relation is often called *concurrence*. We can then say that an object is a mereological fusion of a maximal set of concurring tropes. Furthermore, it is common to construe tropes as non-recurring: no trope can be a member or part of more than one object, and hence two objects cannot have tropes as common members or parts whereby the two objects overlap. Accordingly, concurrence is reflexive, symmetric, and transitive—that is, concurrence is an equivalence relation.[20]

Objects as two-category constructions (substratum theory)

Instead of constructing objects entirely from non-recurring *sui generis* entities, tropes, objects may be constructed partly from recurring *sui generis* entities, which will be called *universals*, together with a further *sui generis* entity that gives the object its particularity. Traditionally, this entity is called a *substratum*.[21] Then we need a primitive relation, a 'fundamental tie', that unites the substratum with the universals. We may call this relation *glue*, and say that an object is the mereological sum of a substratum and all the (monadic) universals glued to that substratum. Since universals recur, two objects may overlap by sharing a universal as a common part.

Other uses of the terms 'trope', 'universal', and 'substratum' and various mixtures of the above positions are possible. My aim here is not to provide a complete list of views, but rather to mention two fairly standard views, in order to show that they are incompatible with the temporal-regions account of temporal supervenience, because they are committed to certain forms of coincidence. I shall consider the two views in turn.

Suppose, first, that an ordinary object is a bundle of tropes—that is, an ordinary object is a mereological fusion of a maximal set of concurring tropes. Suppose that Pia is a particle with mass m at t. According to principle

[19] For what is now regarded as the classic version of trope theory, see Williams (1953). Recent developments of trope theory can be found in Campbell (1990), Simons (1994), and Bacon (1995). See also (Lewis 1986*a*: 1.5).

[20] For a resulting problem about parthood, see Williams (1953: 77) and Bacon (1995: 48–50).

[21] The classic work on universals is Armstrong (1978). See also Lewis (1986*a*: 1.5). The sorts of universals under consideration here are parts of objects, they are 'in' objects, by which feature they have acquired the name *immanent* universals (corresponding to *universalia in rebus*). Immanent universals have been distinguished from *transcendent* universals (corresponding to *universalia ante rem*). These labels derive from Armstrong (1978), who also distinguishes between an Aristotelian and a Platonic conception of universals. Henceforth I shall mean immanent, or Aristotelian, universals when I speak of universals.

(TS_{3D}) of the temporal-regions account, Pia has mass *m* at *t* iff Pia occupies a spacetime region *R* such that *R* is a member of *t* and *R* has mass *m*. So Pia has a spatiotemporal location. Since Pia is the fusion of a maximal set of concurring tropes, Pia's tropes also have a spatiotemporal location. But these tropes are neither spatial nor temporal parts of Pia. They are non-spatiotemporal parts of Pia: the tropes occupy the whole of the spacetime region that Pia occupies. Moreover, (TS_{3D}) entails three-dimensionalism: Pia occupies multiple instantaneous spacetime regions. Since Pia's tropes are spatiotemporally located where Pia is located, Pia's tropes also occupy multiple instantaneous spacetime regions. So we have a case of distinct coincidents: a particle and its tropes all occupy the same spacetime regions.

This form of coincidence leads to trouble. According to (TS_{3D}), something has mass *m* at *t* by occupying a spacetime region that is a member of *t* and that has mass *m*. Since Pia has mass *m* at *t*, Pia occupies a spacetime region *R* such that *R* is a member of *t* and *R* has mass *m*. Since Pia's tropes are located where Pia is located, Pia's tropes also occupy *R*. Moreover, since *R* is a member of *t* and *R* has mass *m*, Pia's tropes have mass *m* at *t*. In general, for every property Pia has at a time *t*, each of the tropes that make up Pia also has this property at *t*, for each of Pia's tropes is where Pia is. This cannot be correct. The kinds of things that have a mass at a time are particles. Tropes are not particles. Hence tropes cannot have a mass at any time. Material objects are therefore not bundles of tropes if the temporal-regions account of temporal supervenience is correct.

Alternatively, a material object may be characterized as a mereological construction from universals together with a substratum that is glued to these universals. Is this a sensible option given (TS_{3D})? Suppose again that Pia is a particle with mass *m* at *t*. By (TS_{3D}), it follows that Pia has a spatiotemporal location. If Pia is a sum of certain universals and a substratum, then both the universals and the substratum also have a spatiotemporal location. Focus on the universals. Analogously to tropes, universals are non-spatiotemporal parts of Pia: the universals occupy the whole of the spacetime region that Pia occupies. Moreover, (TS_{3D}) entails three-dimensionalism: Pia occupies multiple instantaneous spacetime regions. Since Pia's universals are spatiotemporally located where Pia is located, Pia's universals also occupy multiple instantaneous spacetime regions. So we have another case of distinct coincidents: a particle, its substratum, and its universals all occupy the same spacetime regions.

Universals cannot constitute a material object for much the same reason that tropes cannot do so, if (TS_{3D}) is correct. Something has mass *m* at *t* by occupying a spacetime region that is a member of *t* and that has mass *m*. Since Pia has mass *m* at *t*, Pia occupies a spacetime region *R* such that *R* is a member of *t* and *R* has mass *m*. Since Pia's universals are located where Pia is located, Pia's universals also occupy *R*. Moreover, since *R* is a member of *t* and *R* has mass *m*, Pia's universals have mass *m* at *t*. In general, for every property Pia has at a time *t*, each of the universals that make up Pia also has this property at *t*, for each of Pia's universals is where Pia is. The same can be shown for the substratum that partly constitutes Pia, since the substratum is also spatiotemporally located where Pia is located. But neither universals nor substrata are the kinds of things that have a mass at a time. Material objects are therefore not bundles of universals and substrata if the temporal-regions account is correct.

According to the temporal-regions account, a material object is therefore neither a metaphysical construction from non-recurring entities, tropes, nor a construction from recurring entities, universals. This result holds even for fundamental particles that friends of tropes or universals consider as paradigmatic candidates for metaphysically constructed objects. Unless there is a form of construction that does not imply that the constituent entities are spatiotemporally located if the constituted object is so located, the three-dimensionalist is committed to the view that a material object is not a metaphysical construction at all, but rather a *sui generis* kind of entity, sometimes called a *particular*.[22] This outcome does not present an obvious problem for the temporal-regions account; the view of objects as particulars is not obviously inferior to the theories of tropes and universals, and the latter theories are not directly supported by intuition. So, unless the case for believing in the construction of objects from such metaphysical inventions as tropes and universals is strengthened considerably, the three-dimensionalist has little reason to worry. Notice that the result that material objects are not constructed from tropes or universals should not be taken to imply that tropes or universals do not exist. For tropes or universals may be good for constructing other kinds of things that are not subjects of temporal predications.

Coincidence of ordinary objects

What I take to be the most serious threat to the temporal-regions account of temporal supervenience arises from apparent cases in which distinct ordinary objects coincide. I shall focus on the famous case of the statue and the lump of clay.[23] Before looking at this case, a few words on ascriptions of identity are in order.

[22] To say that an object is a particular is not to deny that an object has parts. An object is a *sui generis* entity in that it is not made up from entities that belong to ontological categories distinct from the category it belongs to, although it has constituents belonging to the same category it belongs to—namely, spatial parts and perhaps also temporal parts.

[23] For an overview of other cases of apparent coincidence of distinct ordinary objects, see Sider (2001: 4.1).

Identity is a two-place relation that may be ascribed to objects simpliciter. Identity may also be ascribed to objects at a time. Ascribing identity at a time does not make identity a three-place relation with an extra place for a time. Instead, identity is still two-place, but its ascription is relativized to a time. Many, but not all, ascriptions of identity at the level of ordinary time are relativized in this way.[24] If temporal supervenience holds, then temporally relativized facts of identity are grounded in temporally unrelativized facts of identity. Moreover, the trivial claim that x and y can be identical at a time does not imply the controversial claim that x and y can be identical at one time but distinct at another time. That is, that identity can be temporal—that identity can hold at a time—does not imply that identity can be temporary—that identity can change over time. Finally, given that identity may be atemporal and temporal, we must distinguish between an atemporal and a temporal version of Leibniz's Law (LL). If x is identical to y simpliciter, then, for all properties ϕ, x has ϕ simpliciter iff y has ϕ simpliciter. This is the atemporal version. If x is identical to y at a time t, then, for all properties ϕ, x has ϕ at t iff y has ϕ at t. This is the temporal version.

Suppose now that at time t_1 an artist acquires a lump of clay, and at time t_2 forms a statue using that clay. As a consequence, there is a statue at t_2, and, assuming that the lump has not been destroyed in the artistic process, there is a lump of clay at t_2. Are the lump and the statue identical? Given that identity and distinctness may be relativized to a time, we need to distinguish between the view that the statue and the lump of clay are identical simpliciter, to be called *atemporal monism*, and the view that they are identical at a time, to be called *temporal monism*. Likewise, we need to distinguish between the view that the statue and the lump of clay are distinct simpliciter, to be called *atemporal pluralism*, and the view that they are distinct at a time, to be called *temporal pluralism*. These distinctions are relevant, because it is possible to say coherently that the statue and the lump are distinct simpliciter but identical at a time. A leading proponent of atemporal and temporal pluralism is David Wiggins, who supports his position by claiming that the lump of clay *constitutes* the statue.[25] A friend of the polar opposite position, atemporal and temporal monism, is Michael Burke, who attempts to avoid the coincidence of distinct objects by appealing to the notion of a 'dominant sortal'.[26] As we will see shortly, the

[24] Cross-temporal predications constitute examples of ordinary temporal predications in which the identity-predicate occurs temporally unrelativized; see Sects. 3.3 and 4.5.

[25] See Wiggins (1968; 1980: ch. 1). Other atemporal and temporal pluralists include Doepke (1982), Thomson (1983), Simons (1987), Johnston (1992), Baker (1997), and Fine (2000, 2003).

[26] See Burke (1994).

temporal-regions account is driven towards the combination of atemporal and temporal monism, whereas the temporal-parts account is driven towards the mixed position of atemporal pluralism and temporal monism.

The standard argument for the atemporal distinctness of the statue and the lump of clay rests on an apparent difference between the statue and the lump at t_1: the lump exists at t_1, before the artist goes to work, whereas the statue does not exist at t_1. Since they have different histories, they must be distinct. Here is an explicit statement of the argument:

(A) The lump exists at t_1.
The statue does not exist at t_1.

The lump is distinct from the statue.

(The premisses are listed above the line and the conclusion below.) Suppose now that the lump of clay is identical to the statue. This identity may be rendered compatible with the truth of both premisses of (A) by reading these premisses as saying that one and the same thing is a lump of clay at t_1 but not a statue at t_1. This is the standard reading of the premisses, according to which the singular terms 'the lump' and 'the statue' fall inside the scope of 'at t_1'. On this reading, the conclusion of (A) does not follow from the premisses, since the premisses merely register a difference in the way in which a single object is described, instead of registering different properties of two objects.

What is required in order to make argument (A) work is to give the premisses a different, non-standard reading. Instead of letting the singular terms 'the lump' and 'the statue' fall inside the scope of 'at t_1', the singular terms may be left outside the scope of 'at t_1'. Argument (A) may then be given the following valid reading:

(A′) The lump is such that it exists at t_1.
The statue is such that it does not exist at t_1.

The lump is distinct from the statue.

If the premisses of (A′) are true, then the conclusion stands, since the premisses now register a difference in properties of two objects. While the same object can be a lump at t_1 and yet not be a statue at t_1, the same object cannot both exist at t_1 and fail to exist at t_1. Notice that, since the premisses contain the complex predicate 'exists at t_1', which is satisfied simpliciter, (A′) is an argument for atemporal pluralism from the atemporal version of LL.

Atemporal pluralists may accept the truth of both premisses of (A′). Atemporal monists cannot accept both premisses as true; if the first premiss is true, then the second is false. A rejection of (A′) may be based on the observation that it is far from obvious whether the premisses of (A′) are true. What is obvious is that the unregimented statement that at t_1 the lump but not the statue exists is true. Those who take the statue and the lump to be one may hold that this intuition is captured by the standard reading employed in (A), while pointing out that there is no intuitive ground for the truth of the non-standard reading employed in (A′). The mere observation that the statue and the lump have different histories therefore fails to establish their distinctness.

Consider now a straightforward argument from the temporal version of LL to the distinctness of the statue and the lump at time—namely, at t_2. Instead of appealing to differences between the statue and the lump at t_1, as the traditional argument (A) does, it appeals to differences at t_2. The distinctness of the statue and the lump at t_2 may be established by finding something that is true of the one at t_2 without being true of the other at t_2. Although rarely noticed, there may be many and varied differences between the statue and the lump at t_2. Kit Fine gives the following examples: the statue may be *defective, substandard, well* or *badly made, valuable, ugly, Romanesque, exchanged, insured,* or *admired* at t_2 even though the lump of clay is not.[27] Let us focus on one of these examples and consider the following argument for the distinctness of the statue and the lump at time t_2:

(B) The statue is Romanesque at t_2.
The lump is not Romanesque at t_2.
―――――――――――――――
The statue is distinct from the lump at t_2.

This is an argument for temporal pluralism that works for the standard reading of 'at t_2', which puts the singular terms 'the statue' and 'the lump' inside the scope of 'at t_2' and makes use of the temporal version of LL. Although the traditional dispute arising from arguments such as (A) and (A′) takes place between atemporal monism and atemporal pluralism, I shall focus on the dispute between temporal monism and temporal pluralism arising from arguments such as (B), since (B) is a stronger argument and poses a more serious problem for the temporal-regions account than (A) and (A′).

The problem arising for the temporal-regions account from the possibility of distinct ordinary coincidents, which arguments such as (B) appear to establish, is that the account entails temporal monism. According to (TS_{3D}), *a* is F at *t* iff *a* occupies a spacetime region *R*, such that *R* is a member of *t* &

[27] See Fine (2003: 206).

R is F. Likewise for the dyadic case: a is L to b at t iff a occupies a spacetime region R that is a member of t & b occupies a spacetime region R^* that is a member of t & R is L to R^*. Thus, the statue is identical to the lump at t_2 iff the statue occupies a spacetime region R that is a member of t_2 & the lump occupies a spacetime region R^* that is a member of t_2 & $R = R^*$. It is agreed that the statue and the lump occupy the same place at t_2, and, equivalently, occupy the same instantaneous spacetime region contained in t_2. Hence, the statue and the lump are identical at t_2, which is the temporal monist's stance. By being an argument against temporal monism, argument (B) turns out to be an argument against the temporal-regions account. As a temporal monist, the friend of temporal regions thus faces the burden of blocking argument (B), of showing why argument (B) is invalid.

It is worth mentioning that, while the temporal-regions account is committed to temporal monism, it is not committed to atemporal monism. That the lump of clay and the statue are distinct simpliciter yet occupy the same instantaneous region is not out of the question in a three-dimensionalist world with regional instantiation. What is out of the question is for these distinct coincidents to differ in properties at the time of their coincidence. This is why arguments attempting to establish distinctness simpliciter pose less of a threat than arguments attempting to establish distinctness at a time. However, coincidence of distinct things is commonly considered puzzling. Perhaps there are certain particles that can pass through each other, and hence coincide. But how can distinct ordinary material objects, such as a statue and a lump of clay, fit into the same place at the same time? I consider such overcrowding in the case of ordinary objects worth avoiding. Since the three-dimensionalist can avoid it only by adopting atemporal monism, I shall henceforth portray the three-dimensionalist as a friend of atemporal as well as temporal monism.

In the light of the temporal-regions account's rivalry with the temporal-parts account, it should further be pointed out that argument (B) is just as threatening to the temporal-parts account as it is threatening to the temporal-regions account. To many friends of temporal parts this will come as a surprise, given that it is a commonplace among perdurantists that temporal parts afford an attractive, if not the best, treatment of alleged cases of distinct coincidents.[28] The standard line is that the four-dimensionalist has the resources to accept that the lump and the statue are distinct simpliciter, and to explain why their coincidence at t_2 is unobjectionable. According to the four-dimensionalist, what seems puzzling at the surface is

[28] See, e.g., Sider (2001: 4.2).

at bottom far from puzzling. There really are distinct things, a statue and a lump of clay. But the statue and the lump share their spatial location L at time t_2 by sharing their temporal part at t_2. Since this temporal part is the only thing that strictly occupies location L at t_2, there is no overcrowding. The statue and the lump overlap in L at t_2 without being wholly located in L; they each have temporal parts that do not occupy L at t_2. Such overlap is no more objectionable than overlapping roads. The overlap of the lump and the statue is the temporal analogue of the spatial overlap of a road and one of its subsegments. This is a good explanation of the claim that the statue and the lump are distinct simpliciter yet coincide at a time. But why should we believe this claim? As became clear in the discussion of (A) and (A′), this claim cannot be based on the simple observation that the lump exists at a time at which the statue does not exist. It is doubtful, moreover, whether the kind of coincidence that the friend of temporal parts explains so elegantly derives from any ordinary observation at all.

The real puzzle of coincidence is that there seems to be compelling evidence for the claim that objects that are distinct *at a time* can coincide at that time. In the case of the statue and the lump of clay, the latter kind of coincidence is apparently established by argument (B). While the conclusion of (A) and (A′) is readily explained by the theory of temporal parts, the conclusion of (B) is incompatible with the theory of temporal parts, given that the theory is committed to (TS_{4D}). According to (TS_{4D}), a is R to b at t iff a has a temporal part, a_t, that occupies a member-region of t & b has a temporal part, b_t, that also occupies a member-region of t & a_t is R to b_t. Since the temporal part of the statue that occupies a member-region of t_2 is identical to the temporal part of the lump that occupies a member-region of t_2, it follows that the statue and the lump are identical at t_2, which is an instance of temporal monism and which contradicts the conclusion of (B). It turns out that (B) is an argument against the temporal-parts account as much as it is an argument against the temporal-regions account, which deprives the four-dimensionalist of what is often viewed as an important advantage over the three-dimensionalist.

In the remainder of this section, I will do three things. First, I will show by what move argument (B) can be blocked—that is, I will show how to get temporal monism, and hence the temporal-regions account (as well as the temporal-parts account) off the hook. Secondly, I will defend this temporal monist response against an important objection. And, thirdly, I will give an argument against temporal pluralism that is closely related to the temporal monist response. (Henceforth, I will mean temporal monism and temporal pluralism when I speak of monism and pluralism, respectively.)

Monism and sortal relativity

Coincidence has so far been understood in the sense of *spatial* coincidence at a time. Two things spatially coincide at a time when the exact place that they occupy at that time is the same. In the three-dimensionalist framework, spatial coincidence at a time becomes instantaneous spatiotemporal coincidence—that is, sameness of instantaneous spacetime region occupied. Alternatively, coincidence may be understood as *material* coincidence at a time. Two things materially coincide at a time when their underlying matter at that time is the same.[29] If things can materially coincide at a time without spatially coinciding at that time, then material coincidence at t does not entail identity at t, according to (TS_{3D}), whereas spatial coincidence at t does entail identity at t. The proponent of (TS_{3D}) could then adopt pluralism by claiming that all alleged cases of spatial coincidence are really cases of purely material coincidence. This line of response is implausible. I find it hard to imagine a case in which things share their matter at t without sharing their spatial location at t. And, even if there were such cases, the case of the statue and the lump of clay is not one of them. The statue and the lump of clay share the same underlying matter at t_2 as well as the same place at t_2. In what follows, I will argue against coincidence at a time understood as spatial and material coincidence.

So where does argument (B) go wrong? Fine discusses a number of options of explaining why arguments such as (B) break down and concludes that 'predicational shift' is the only viable option.[30] I agree with Fine that the monist is committed to predicational shift. But I will not argue for this or repeat what Fine says. For, in order to get the monist off the hook, it will suffice to find one good way of avoiding the threat of coincidence, and I will show that predicational shift is such a way. (Note, however, that the monist's commitment to predicational shift will become relevant in the argument against pluralism to be given below.)

The intuitive idea behind predicational shift is that, in asserting that the statue is Romanesque and that the lump of clay is not, there must be a shift in the property attributed to the same object. Such a shift is triggered by how the object is conceived of. We may conceive of a given object as a statue or as a lump of clay, and, depending upon how it is conceived, we will be willing to say one thing about it rather than another. This idea may be cashed out by letting the premisses of (B) contain an implicit modifier

[29] For the distinction between spatial and material coincidence, see also Fine (2003: 198).

[30] See Fine (2003: 208–11).

of the form '*as an S*', where '*S*' is some sortal term—for instance, 'statue' or 'lump'. 'The statue is Romanesque at *t*' is thus understood as 'The statue is Romanesque *as a statue* at *t*' and 'The lump is not Romanesque at *t*' is understood as 'The lump is not Romanesque *as a lump* at *t*'. If the premisses of (B) are sortally relativized in this way, then the conclusion does not stand. For the statue may be identical to the lump at *t* and yet be Romanesque as a statue without being Romanesque as a lump.

If this predicational-shift gambit is to be available to the three-dimensionalist, then the temporal-regions account of temporal supervenience must be extended to cover sortally relativized facts. Principle (TS_{3D}) of the temporal-regions account links ordinary temporal facts with spacetime facts. The following extension of (TS_{3D}) links ordinary sortally relativized temporal facts with spacetime facts: necessarily, *a* is F *as an S* at *t* iff *a* occupies a spacetime region that is a member of *t* and that is F *as an S* simpliciter.[31]

Is sortal relativity anomalous?

Regarding sortal relativity, Fine points out that the fact that *if* the premisses of (B) are sortally relativized then (B) is invalid has no bearing on the question whether the argument really is invalid, unless the sortal-relativization strategy faithfully reflects our actual use of language. With this condition in mind, Fine argues against the monist response to (B), accumulating evidence that is meant to show that sortal relativity 'has intolerable consequences for the functioning of our language' (2003: 198). Since a full response to his criticism would require its own chapter, I shall rest content with replying to one of his objections, which I consider fundamental, and which motivates a refinement of sortal relativity as sketched above.

[31] Similarly, (TS_{4D}) must be extended to allow for sortally relativized temporal predications: necessarily, *a* is F *as an S* at *t* iff *a* has a temporal part that occupies a member-region of *t* and that is F *as an S* simpliciter. We encountered this principle as (TS^{**}_{4D}) in 4.5, and distinguished it from (TS^{*}_{4D}): necessarily, *a* is F *as an S* at *t* iff *a* is an *S* simpliciter, and *a* has a temporal part that occupies a member-region of *t* and that is F simpliciter. We also saw in Sect. 4.5 that, if (TS^{*}_{4D}) is correct, then the modifier '*as an S*' is detachable: if the statue is Romanesque as a statue at *t*, then the statue is Romanesque at *t*; and, if the lump is not Romanesque as a lump at *t*, then the lump is not Romanesque at *t*. This means that, if predicational shift via sortal relativity is indeed the only viable option for the four-dimensionalist to block argument (B), then (TS^{*}_{4D}) must be rejected in favour of (TS^{**}_{4D}), or else sortal relativity becomes ineffective. The same point can be made against the three-dimensionalist analogue of (TS^{*}_{4D}), which construes sortal modifiers as detachable: necessarily, *a* is F *as an S* at *t* iff *a* is an *S* simpliciter, and *a* occupies a region that is a member of *t* and that is F simpliciter.

The objection is that sortally relativized predicates do not conform with one's general understanding of how respect-relative predicates should behave.[32] This can be shown by comparison with predicates that, according to Fine, are clearly respect relative and do conform to these general principles. Among the principles Fine has in mind is the following: it should be possible to specify the relevant respect to which a given predicate is relativized (*a*) by explicit qualification of the predicate-term, (*b*) by the subject-term, or (*c*) by means of the context. The predicate 'is qualified' straightforwardly meets these general criteria for respect relativity. As regards criterion (*a*), it is natural to say that Sam is qualified *for the position of janitor*. The predicate 'is qualified' is true of Sam relative to the respect 'position of janitor', which can be indicated within the predicate-term itself by adding 'for the position of janitor'. As Fine puts it, by indicating the respect within the predicate 'one brings it "home" to where it most naturally belongs'.[33] As regards criterion (*b*), we may say that the person who applied for the position of janitor is qualified. Here the relevant respect is specified by the subject-term 'the person who applied for the position of janitor'. Given that the respect is primarily associated with the predicate, as pointed out under (*a*), this second example may be read as elliptical for 'The person who applied for the position of janitor is qualified for the position of janitor', where for reasons of economy the second occurrence of 'the position of janitor' may be dropped because it already governs the subject-term. Moreover, it makes sense to say that the person who applied for the position of professor is qualified for the position of janitor. Here the subject-term, although it contains a certain respect, plays no role in picking out the relevant respect. That is, the respect in the predicate-term trumps the respect in the subject-term. A related case of trumping is at work in cases satisfying criterion (*c*). Suppose, following Fine, that in a desperate attempt to find qualified candidates for the position of professor we search through the candidates for other positions and appropriately say that the person who applied for the position of janitor is qualified, meaning that he is qualified for the position of professor. In this case the respect that is introduced by the non-linguistic context, and that may be explicitly indicated within the predicate-term, trumps the respect indicated by the subject-term.

Are criteria (*a*)–(*c*) satisfied when the respect to which a given predicate is relativized is a sort? Fine says 'no'. The main problem is posed by criterion (*a*). For to say that something is Romanesque *as a statue*, or *qua statue*, does not receive any sanction from ordinary use. The phrases 'as a statue' and 'qua statue' are phrases of art, 'philosophical inventions [without] basis in

[32] See Fine (2003: 214). [33] ibid.

ordinary usage'.[34] Thus, says Fine, a sort can never be specified through explicit qualification of the predicate itself, and hence criterion (*a*) cannot be satisfied. The relevant respect is therefore not primarily associated with the predicate-term, as in the case of 'is qualified', but rather with the subject-term. And so all the burden of indicating the relevant sort lies on the subject-term. When we say that the statue is Romanesque, the predicate 'is Romanesque' is relativized to the sort invoked by the subject-term 'the statue'. As a consequence, it is unlikely that sortal relativity permits a case of contextual trumping satisfying criterion (*c*), in which a sort specified by the non-linguistic context overrides the sort specified by the subject-term. (Note, however, that the context may specify a sort in cases where the subject-term is not governed by a sort, such as 'Val is Romanesque'.) The objection to sortal relativity, then, is that it fails to satisfy criteria (*a*) and (*c*), and hence that sortal relativity is problematically anomalous.

Consider the sortally relativized predication '*a* is F as an *S*', where '*S*' is some sortal term. I shall now state my account of sortal predications of this form, and show that on this account sortal relativity is not anomalous, in virtue of satisfying criteria (*a*)–(*c*). My proposal is to treat '*a* is F as an *S*' as a complicated way of saying that *a* is an F*S*. Thus, 'Val is Romanesque as a statue' becomes 'Val is a Romanesque statue'. The adjective 'F' in '*a* is an F*S*' may function either as a *predicative* adjective or as an *attributive* adjective. If 'F' functions as a predicative adjective, then '*a* is an F*S*' implies that *a* is F; otherwise 'F' functions as an attributive adjective. Standard cases of predicative constructions are 'round ball' and 'black record'; standard cases of attributive constructions are 'brilliant dancer' and 'crazy driver'. A black record is black and a round ball round; but a crazy driver need not be crazy, nor a brilliant dancer brilliant. In the attributive case, the adjective does not form a predicate with the copula 'is', but rather functions as a modifier of a sortal noun, and is semantically dependent on this noun. Given the predicative–attributive distinction, my hypothesis is that 'Romanesque' in 'Val is a Romanesque statue' functions as an attributive adjective.

A semantics of sortal predicates of the form 'is an F*S*', where 'F' is an attributive adjective, may be sketched as follows: 'F' is associated with a function that takes the extension of 'is an *S*' as its argument. In the standard case, 'F' is associated with a function from the extension of 'is an *S*' to a subset of this extension. Accordingly, the extension of 'Romanesque statue' is a subset of the extension of 'statue'. This has the desirable consequence that standard attributives are detachable: if Val is a Romanesque statue, then Val is a statue. But the same does not hold for 'Romanesque', for nothing is

[34] ibid. (2003: 215).

plain Romanesque. The adjective 'Romanesque' functions as a modifier of a sortal term, and as such has no extension on its own. I will call this account of sortal predication the *sortal-modifier account*. To emphasize, the need for attributive adjectives has widely been recognized to arise from such cases as 'brilliant dancer' and 'crazy driver'. So the sortal-modifier account does not introduce an altogether new category of adjectives. It rather extends the class of attributive adjectives to include adjectives that have hitherto been viewed as predicative, such as 'Romanesque' in 'Val is a Romanesque statue'.

It must further be pointed out that the sortal-modifier account of sortal predication can be straightforwardly incorporated into (TS_{3D}), the core principle of the temporal-regions account of temporal supervenience: necessarily, *a* is an F*S* at *t* iff *a* occupies a spacetime region that is a member of *t* and that is an F*S* simpliciter.[35] This point is obviously important, since it is not merely the monist whom we want to free from the pluralist charge, but the monist who holds the temporal-regions account.

Now back to criteria (*a*)–(*c*). As regards (*a*), the sortal-modifier account has the consequence that, if sortal predications are brought into the correct form—that is, if 'Val is Romanesque as a statue' is read as 'Val is a Romanesque statue'—then we end up with an entirely common way of indicating the sort to which the predicate is relativized within the predicate itself, a way of putting the sortal 'where it belongs'. This result also provides a treatment of predications satisfying criterion (*b*)—predications in which the relevant sort is indicated by the subject-term—that is analogous to the above treatment of 'The person who applied for the position of janitor is qualified': 'The statue is Romanesque' may be read as elliptical for 'The statue is a Romanesque statue'. The sortal information in the predicate may be dropped for reasons of economy, because it is already given by the sortal governing the subject-term. The ellided part may therefore be unambiguously specified given the linguistic context. This is exactly how the premisses of argument (B) should be understood. Furthermore, it makes sense to say that the lump of clay is a Romanesque statue. Here the subject-term, although it invokes a certain sort, plays no role in picking out the relevant sort. The sort invoked by the predicate-term trumps the sort invoked by the subject-term. Cases satisfying criterion (*c*) are then not far away. Imagine that we are in an exhibition of 1960s Conceptual Art. In this non-linguistic context it may be appropriate to say that the piece of metal in the corner is pretentious, meaning that it is a pretentious work of art. Here the sort that is introduced by the non-linguistic context, and

[35] Analogously for (TS_{4D}): necessarily, *a* is an F*S* at *t* iff *a* has a temporal part that occupies a member-region of *t* and that is an F*S* simpliciter.

that may be explicitly indicated within the predicate-term, trumps the sort invoked by the subject-term. All of these features of sortally relativized predicates are perfectly analogous to the features of 'is qualified' pointed out earlier. Thus, sortally relativized predication when construed along the lines of the sortal-modifier account satisfies criteria (*a*)–(*c*) and therefore is far from anomalous. Instead, the above considerations suggest that the sortal-modifier account is a natural hypothesis about the functioning of our language, and hence a workable explanation for the proponent of the temporal-regions account of temporal supervenience to adopt in order to avoid the threat of distinct coincidents.[36]

Pluralism and sortal relativity

The monist may invoke predicational shift via sortal relativity to counter argument (B). The pluralist, on the other hand, denies that any predicational shift occurs in argument (B), and asserts that the premisses are free of sortal relativity. In what follows, I shall strengthen the case for monism further by criticizing this pluralist attitude in the light of Fine's view that predicational shift via sortal relativity is the only viable option of breaking arguments such as (B).

The first step in my critique of pluralism is to extend the case of the statue and the lump of clay. Consider how the story about the artist might continue. Suppose that, after having moulded the clay into a statue, and after having failed to sell the statue, the artist hands over the statue to his children to play with as a toy. And one night he even uses the statue as a weapon to threaten a burglar. The sortals that occur in this story are 'lump of clay', 'statue', 'toy', and 'weapon'. The difference between how 'lump of clay' and 'statue' occur and how 'toy' and 'weapon' occur may be put as follows. In the story about the artist an object is *created* as a lump of clay and an object is *created* as a statue, but an object is merely *used* as a toy and as a weapon.[37]

Why is an object *merely* used as a toy and as a weapon? Could we not say that, when the statue is used as a toy, a toy is created, in which case an object would be created as an *S* by using another object as an *S*? If an object is created through use, then the object's existence is dependent on

[36] An alternative response to Fine's anomaly charge is developed by Jeffrey C. King in an unpublished paper entitled 'Semantics for Monists'.

[37] Some monists will deny that in the story about the artist an object is created as a statue. They will say, instead, that an already existent object, the lump of clay, becomes a statue, and hence will deny that 'statue' here functions as a substance sortal (as defined below). This point may be neglected, because the focus of the present discussion is on how the pluralists view the story.

use. If the toy is created as an object distinct from the statue by using the statue as a toy, then the toy exists only if someone uses the statue as a toy. The toy comes into existence when the statue is first used as a toy, and the toy ceases to exist as soon as the statue ceases to be used as a toy. This use-dependent existence, however, does not seem to be a mark of material objects. Material objects can exist 'left alone', independently of the ways in which we use them. This is why it is not sensible to claim that by using the statue as a toy a toy is created, assuming that the toy is a material object.

A further important point concerning the distinction between creating something as an *S* and using something as an *S*, for some sortal term '*S*', is this: whether an object *x* is created as an *S*, or whether *x* is merely used as an *S*, *x* can be correctly described as an *S*—that is, sortal '*S*' applies to *x*. For an artefact *x* can be described as an *S* in virtue of functioning as an *S*, whether *x* functions as an *S* because it was created as an *S* or because it is merely used as an *S*. For example, it makes sense to say that the statue is the kids' favourite toy now, in which case something is described as a toy on the basis of merely being used as a toy and not having been created as a toy. Moreover, there are cases in which we describe an object *x* as an *S* on the basis that *x* functions as an *S*, but without knowing whether *x* was created as an *S* or whether *x* is merely used as an *S*. We have such a case when the police search for the weapon by means of which the burglar was killed, and later identify the statue as that weapon.

The intuitive distinction between being created as an *S* and merely being used as an *S* is closely related to the distinction between *substance sortals* and *phase sortals*.[38] Here are fairly standard definitions of these notions:

> '*S*' applies to *x* as a *substance sortal* $=_{\mathrm{df}}$ '*S*' applies to *x* at all times and at all possible worlds at which *x* exists.
> '*S*' applies to *x* as a *phase sortal* $=_{\mathrm{df}}$ '*S*' applies to *x* only at some times or only at some possible worlds at which *x* exists.[39]

These definitions may be linked with the notions of being created as an *S* and being merely used as an *S* by means of the following conditionals: if *x* is *created* as an *S*, then '*S*' applies to *x* at all times and at all possible worlds at which *x* exists; whereas if *x* is merely *used* as an *S*, then '*S*' applies to *x* only at some times or only at some possible worlds at which *x* exists. These conditionals presuppose what was argued in the preceding paragraph: that a sortal '*S*' may apply to *x* because *x* was created as an *S* or because *x* is

[38] See Wiggins (2001: 30).

[39] The right-hand side of this definition does not read '"*S*" applies to *x* only at some times *and* only at some possible worlds at which *x* exists', because a phase sortal may accidentally apply to *x* at all times of its existence.

merely used as an S. Thus, if x is created as an S, then 'S' applies to x as a substance sortal; and, if x is merely used as an S, then 'S' applies to x as a phase sortal. In the story about the artist as the pluralists conceive of it, 'lump of clay' and 'statue' function as substance sortals, whereas 'toy' and 'weapon' function as phase sortals. However, 'toy' and 'weapon' do not always function as phase sortals. For example, 'toy' applies to a Playstation as a substance sortal. This is why the definitions of substance sortal and phase sortal are sensitive to which object the sortal applies to.

Let us now return to argument (B) for distinct coincidents. The argument rests on predicational differences between the lump of clay and the statue at some time. Taking into account the extended story about the artist, there are times at which there is a toy and there are times at which there is a weapon, in addition to the statue and the lump that exist at all those times. Furthermore, just as there are predicational differences between the lump and the statue, there are predicational differences between the lump, the statue, the toy, and the weapon. For example, the toy may be dangerous and the weapon deadly, but we would be surprised to find a dangerous statue or a deadly lump of clay; the lump of clay may be chemically impure, but not the toy; and the toy may be well or badly made, though not the lump of clay. Suppose further that I use my alarm clock as a paperweight. Then there is a time at which there is an alarm clock and a paperweight. At this time it may be true to say that the alarm clock is radio controlled, but it would hardly be sensible to speak of a radio-controlled paperweight.[40]

By Leibniz's Law, it follows that at some time or other the toy and the weapon are distinct from the statue and from the lump of clay. Likewise, it follows that there is a time at which the alarm clock is distinct from the paperweight. In order to focus the discussion, let us fix on the following sample argument:

(C) The toy is dangerous at t.
The statue is not dangerous at t.

The toy is distinct from the statue at t.

[40] When assessing these examples of predicational differences it is important to keep in mind the dialectical context of the present discussion. The examples are directed against the pluralist of Finean stripe who accepts, for example, that the statue and the lump may differ in that the one is well or badly made though not the other. Such a pluralist should also accept that the toy and the lump may differ in that the one is well or badly made though not the other. Note further that there are also predicational differences between the toy and the weapon. But these predicational differences are less important than those between the toy, the statue, and the lump, because in our story the toy and the weapon exist at difference times, and hence there is no chance of running an argument for their distinctness at a time from Leibniz's Law.

Not even a hard-nosed pluralist will want to live with the conclusion of (C). It is sensible to say that, when the artist moulds the clay, a new thing, a statue, comes into existence. However, it is not sensible to say that, when the artist hands over the statue to his children to play with, a new thing, a toy, comes into existence. As pointed out above, the existence of a use-created object is use-dependent, but the existence of material objects, such as the toy in our story, is not use-dependent. Instead of a new thing being created as a toy, an already existent thing, the statue, is merely used as a toy, which implies that the statue is the toy. For analogous reasons it would not be sensible to distinguish between the toy and the lump of clay, between the weapon and the statue or the lump of clay, and between the paperweight and the alarm clock, despite the mentioned predicational differences.

(B) is an argument for the coincidence of distinct things both of which fall under a substance sortal, whereas (C) is an argument for the coincidence of distinct things one of which falls under a phase sortal. While the status of the conclusion of (B) is controversial, the status of the conclusion of (C) is not: the conclusion of (C) is false. The pluralist therefore finds herself in the position of endorsing argument (B) while rejecting argument (C). The resulting problem for the pluralist is to explain this asymmetry, to explain why argument (B) goes through and argument (C) does not.

The pluralist might be tempted to explain the asymmetry by appealing to the distinction between substance sortals and phase sortals. But this distinction does nothing to explain why (B) goes through and (C) does not. This is so because the source of the predicational differences that power (B) and (C) is analogous in both cases. The predicational difference in (C) as well as the one in (B) lies, as Fine puts it, in the *correct* and *meaningful* application of a predicate.[41] A statue can meaningfully be said to be Romanesque, though not a lump of clay. A toy can meaningfully be said to be dangerous, though not a statue (at least not in the same literal sense). Or, to use one of Fine's sample predicates, a toy can meaningfully be said to be well made, but not a lump of clay. Likewise, a weapon can meaningfully be said to be deadly, but neither a lump of clay nor a statue, while a lump of clay can meaningfully be said to be chemically impure, but not a toy. In general, both substance sortals and phase sortals may be associated a characteristic range of predicates that have meaningful application to objects in the extension of the sortal. To speak with Fine, both substance sortals and phase sortals have 'spheres of discourse'.[42] Accordingly, the predicational differences in (B) and (C) have an analogous source irrespective of which kind of sortal is employed: the predicate

[41] Fine (2003: 207). [42] ibid.

employed in (B) and (C) respectively falls in the sphere of discourse of the sortal in the first premiss, but does not fall in the sphere of discourse of the sortal in the second premiss. In whichever way we characterize the difference, in the story about the artist, between the use of the sortals 'statue' and 'lump of clay' and the use of the sortals 'toy' and 'weapon', the fact remains that all these sortals have different spheres of discourse. And it is this feature of sortal nouns, a feature that cuts across the distinction between substance and phase sortals, that raises the problem for the pluralist, because this feature together with Leibniz's Law is responsible for the appearance of distinct coincidents. (Notice here that there are many predicational differences, apparently sufficient to establish distinctness, that lie merely in the correct application of a predicate, but not in the meaningful application—that is, predicational differences where the predicate falls in the sphere of discourse of both sortals involved. It may, for example, be correct to say that the statue is boring, and false but still meaningful to say that the lump or the toy is boring. Likewise, it may be correct to say that the alarm clock is broken, and false but meaningful to say that the paperweight is broken.)

The pluralist is committed to the view that (B) is valid and (C) is not. In other words, the pluralist about (B) must be a monist about (C). I doubt that this asymmetry can be sustained. Given that the predicational differences in (B) and (C) have an analogous source, my suspicion is that any strategy that may be invoked to explain why (C) breaks down can equally be invoked to explain why (B) breaks down, and hence that there is no way for the pluralist to explain the asymmetry between (B) and (C). There is no room, however, to pursue this hypothesis for pluralism in general. Let us therefore focus on the pluralist of Finean persuasion, the pluralist who believes that the only viable way of blocking arguments like (B) is to appeal to predicational shift triggered by sortal relativity. Since the predicational difference in (C), just as the predicational difference in (B), is rooted in differing spheres of discourse associated with the sortals 'statue' and 'toy', respectively, the Finean pluralist is committed to blocking (C) by invoking sortal relativity, where the relativizing sortal may be a substance sortal or a phase sortal. The first premiss of (C) may then be understood as 'The toy is dangerous *as a toy* at *t*'—or, presupposing the sortal-modifier account, 'The toy is a dangerous toy at *t*'—and the second premiss may be understood as 'The statue is not dangerous *as a statue* at *t*'—or, presupposing the sortal-modifier account, 'The statue is not a dangerous statue at *t*'. If the premisses of (C) are sortally modified in this way, then the conclusion does not stand. For the toy may be identical to the statue at *t* and yet be dangerous as a toy without being dangerous as a statue. By choosing this route, the Finean pluralist commits herself to the presence in (C) of exactly the kind of sortal relativity that she wants to keep out of (B). Since the predicational difference in (C) shares its source with the predicational difference in (B), it is natural to expect that, if the sortal-relativization strategy applies to (C), then it should apply to (B) as well, which would contradict the pluralist claim that (B) is free of sortal relativity.

Let me draw a conclusion. The puzzles of coincidence take us to a junction where we have to decide whether our *world* is really more complex than we thought—which is the pluralist route—or whether our *language* is really more complex than we thought—which is the monist route. I have taken some steps towards showing that the monist route is a promising one, and that what makes our language more complex than we thought is the presence of sortal relativity. I have further shown that even the pluralist needs the apparatus of sortal relativity to handle certain arguments for distinct coincidents involving phase sortals. But this very apparatus undermines the pluralist's treatment of arguments for distinct coincidents involving only substance sortals, a treatment according to which such arguments are free of sortal relativity. This defence of monism was undertaken in the service of the temporal-regions account of temporal supervenience, which rules out qualitatively different coincidents. As far as statues and lumps of clay are concerned, the threat of coincidence is banned.

References

Adams, Robert Merrihew (1986). 'Time and Thisness', in P. A. French, T. E. Uehling Jr., and H. K. Wettstein (eds.), *Midwest Studies in Philosophy*, xi. *Studies in Essentialism*. Minneapolis: University of Minnesota Press, 315–29.

Armstrong, D. M. (1978). *Universals and Scientific Realism, i. Nominalism and Realism; ii. A Theory of Universals*. Cambridge: Cambridge University Press.

—— (1980). 'Identity through Time', in P. van Inwagen (ed.), *Time and Cause: Essays Presented to Richard Taylor*. Dordrecht: D. Reidel, 67–78.

Augustine (1948). *Basic Writings of Saint Augustine*. New York: Random House.

Bacon, John (1995). *Universals and Property Instances: The Alphabet of Being*. Oxford: Blackwell.

Baker, Lynne Rudder (1997). 'Why Constitution is not Identity', *Journal of Philosophy*, 94: 599–621.

Balashov, Yuri (1999). 'Relativistic Objects', *Noûs*, 33: 644–62.

—— (2000). 'Enduring and Perduring Objects in Minkowski Space–Time', *Philosophical Studies*, 99: 129–66.

Balashov, Yuri, and Janssen, Michel (2003). 'Presentism and Relativity', *British Journal of the Philosophy of Science*, 54: 327–46.

Barker, Stephen, and Dowe, Phil (2003). 'Paradoxes of Multi-Location', *Analysis*, 63: 106–14.

—— (2005). 'Endurance is Paradoxical', *Analysis*, 65: 69–74.

Bennett, Jonathan (1988). *Events and their Names*. Indianapolis: Hackett.

Bigelow, John (1996). 'Presentism and Properties', in J. E. Tomberlin (ed.), *Philosophical Perspectives*, x. *Metaphysics* (Cambridge, MA: Blackwell), 35–52.

Broad, C. D. (1923). *Scientific Thought*. New York: Harcourt, Brace, & Company.

Burke, Michael (1994). 'Preserving the Principle of One Object to a Place: A Novel Account of the Relations among Objects, Sorts, Sortals, and Persistence Conditions', *Philosophy and Phenomenological Research*, 54: 591–624.

Butterfield, Jeremy (1984*a*). 'Spatial and Temporal Parts', *Philosophical Quarterly*, 35: 32–44.

—— (1984b). 'Seeing the Present', *Mind*, 93: 161–76. Reprinted in Robin LePoidevin (ed.), *Questions of Time and Tense* (Oxford: Oxford University Press, 1998), 61–76.

Callendar, Craig (2001). 'Humean Supervenience and Rotating Homogeneous Matter', *Mind*, 101: 25–44.

Campbell, Keith (1990). *Abstract Particulars*. Oxford: Blackwell.

Carter, William, and Hestevold, H. S. (1994). 'On Passage and Persistence', *American Philosophical Quarterly*, 31: 269–83.

Casati, Roberto, and Varzi, Achille C. (1999). *Parts and Places: The Structures of Spatial Representation*. Cambridge, MA: MIT Press.

Castañeda, Hector-Neri (1967). 'Indicators and Quasi-Indicators', *American Philosophical Quarterly*, 4(2): 85–100.

Chalmers, David (1996). *The Conscious Mind: In Search of a Fundamental Theory*. Oxford: Oxford University Press.

Charles, David (2000). *Aristotle on Meaning and Essence*. Oxford: Oxford University Press.

Chisholm, Roderick (1976). *Person and Object: A Metaphysical Study*. La Salle, IL: Open Court.

Craig, William Lane (2000*a*). *The Tensed Theory of Time: A Critical Examination*. Dordrecht: Kluwer Academic Publishers.

——(2000*b*). *The Tenseless Theory of Time: A Critical Examination*. Dordrecht: Kluwer Academic Publishers.

——(2001). *Time and the Metaphysics of Relativity*. Dordrecht: Kluwer Academic Publishers.

Crisp, Thomas M. (2003). 'Presentism', in Michael J. Loux and Dean W. Zimmerman (eds.), *The Oxford Handbook of Metaphysics*. Oxford: Oxford University Press, 211–45.

——(2004*a*). 'On Presentism and Triviality', in D. Zimmerman (ed.), *Oxford Studies in Metaphysics*. Oxford: Oxford University Press i., 15–20.

——(2004*b*). 'Reply to Ludlow', in D. Zimmerman (ed.), *Oxford Studies in Metaphysics*. Oxford: Oxford University Press i., 37–46.

Davidson, Donald (1967). 'Truth and Meaning', *Synthese*, 17: 304–23.

Divers, John (1999). 'A Genuine Realist Theory of Advanced Modalizing', *Mind*, 108: 217–39.

Doepke, Frederick (1982). 'Spatially Coinciding Objects', *Ratio*, 24: 45–60.

Dyke, Heather (1998). 'Real Times and Possible Worlds', in Robin LePoidevin (ed.), *Questions of Time and Tense*. Oxford: Oxford University Press, 93–118.

Earman, John (1989). *World Enough and Space–Time: Absolute versus Relational Theories of Space and Time*. Cambridge, MA: MIT Press.

Fine, Kit (1985). 'Plantinga on the Reduction of Possibilist Discourse', in J. E. Tomberlin and P. van Inwagen (eds.), *Alvin Plantinga*. Dordrecht: D. Reidel, 145–86.

——(2000). 'A Counter-Example to Locke's Thesis', *Monist*, 83: 357–61.

——(2003). 'The Non-Identity of a Material Thing and its Matter', *Mind*, 112: 198–234.

Forbes, Graeme (1985). *The Metaphysics of Modality*. Oxford: Oxford University Press.

——(1987). 'Is There a Problem about Persistence?', *Aristotelian Society Supplementary Volume* 61: 137–55.

——(1989). *Languages of Possibility: An Essay in Philosophical Logic*. London: Blackwell.

Frege, Gottlob (1918/1956). 'The Thought', *Mind*, 65: 289–311.

Gale, Richard (1968). *The Language of Time*. London: Routledge & Kegan Paul.

Gallois, André (1998). *Occasions of Identity*. Oxford: Oxford University Press.

Geach, Peter T. (1980). *Reference and Generality*, 3rd edn. Ithaca, NY: Cornell University Press.

Gibbard, Allan (1975). 'Contingent Identity', *Journal of Philosophical Logic*, 4: 187–221.

Giorgi, Alessandra, and Pianesi, Fabio (1997). *Tense and Aspect*. Oxford: Oxford University Press.

Goodman, Nelson (1951). *The Structure of Appearance*. Cambridge, MA: Harvard University Press.

Haslanger, Sally (1989*a*). 'Endurance and Temporary Intrinsics', *Analysis*, 49: 119–25.

——(1989*b*). 'Persistence, Change, and Explanation', *Philosophical Studies*, 56: 1–28.

Hawley, Katherine (1998). 'Why Temporary Properties Are Not Relations between Objects and Times', *Proceedings of the Aristotelian Society*, 98: 211–16.

——(1999). 'Persistence and Non-Supervenient Relations', *Mind*, 108: 53–67.

——(2001). *How Things Persist*. Oxford: Oxford University Press.

Heller, Mark (1984). 'Temporal Parts of Four Dimensional Objects', *Philosophical Studies*, 46: 323–34.

——(1990). *The Ontology of Physical Objects: Four Dimensional Hunks of Matter*. Cambridge: Cambridge University Press.

——(1992). 'Things Change', *Philosophy and Phenomenological Research*, 52: 695–704.

——(1993). 'Varieties of Four-Dimensionalism', *Australasian Journal of Philosophy*, 71: 47–59.

Hinchliff, Mark (1996). 'The Puzzle of Change', in J. Tomberlin (ed.), *Philosophical Perspectives*, x. *Metaphysics*. Cambridge, MA: Blackwell, 119–36.

Hirsch, Eli (1982). *The Concept of Identity*. Oxford: Oxford University Press.

Hornstein, Norbert (1990). *As Time Goes By*. Cambridge, MA: MIT Press.

Hudson, Hud (1999). 'Temporal Parts and Moral Personhood', *Philosophical Studies*, 93: 299–316.

——(2001). *A Materialist Metaphysics of the Human Person*. Ithaca, NY: Cornell University Press.

Johnston, Mark (1987). 'Is There a Problem about Persistence?', *Aristotelian Society Supplementary Volume*, 61: 107–35.

——(1992). 'Constitution is Not Identity', *Mind*, 101: 89–105.

Kaplan, David (1979). 'On the Logic of Demonstratives', in P. A. French, T. E. Uehling Jr., and H. K. Wettstein (eds.), *Contemporary Perspectives in the Philosophy of Language*. Minneapolis: University of Minnesota Press, 401–12.

——(1989). 'Demonstratives', in J. Almog, J. Perry, and H. Wettstein (eds.), *Themes from Kaplan*. New York: Oxford University Press, 481–563.

Kim, Jaegwon (1982). 'Psychophysical Supervenience', *Philosophical Studies*, 41: 51–70.

——(1984). 'Concepts of Supervenience', *Philosophy and Phenomenological Research*, 45: 153–76.

Kripke, Saul (1977). 'Speaker's Reference and Semantic Reference', in P. A. French, T. E. Uehling Jr., and H. K. Wettstein (eds.), *Studies in the Philosophy of Language. Midwest Studies in Philosophy*, ii. Minneapolis: University of Minnesota Press, 6–27.

Langton, Rae, and Lewis, David (1998). 'Defining "Intrinsic"', *Philosophy and Phenomenological Research*, 58: 333–45.

LePoidevin, Robin (1991). *Change, Cause and Contradiction*. New York: St Martin's Press.

Lewis, David (1968). 'Counterpart Theory and Quantified Modal Logic', *Journal of Philosophy*, 65: 113–26.

——(1971). 'Counterparts of Persons and their Bodies', *Journal of Philosophy*, 68: 203–11.

——(1976). 'Survival and Identity', in A. Rorty (ed.), *The Identities of Persons* (Berkeley and Los Angeles: University of California Press), 17–40. Reprinted with postscript in his *Philosophical Papers* (Oxford: Oxford University Press, 1983), i. 55–77.

——(1979). 'Attitudes *De Dicto* and *De Se*', *Philosophical Review*, 88: 513–43.

——(1983). 'New Work for a Theory of Universals', *Australasian Journal of Philosophy*, 61: 343–77. Reprinted in Lewis (1999*a*), 8–55.

——(1986*a*). *On the Plurality of Worlds*. Oxford: Blackwell.

——(1986*b*). *Philosophical Papers*, ii. Oxford: Oxford University Press.

——(1988). 'Rearrangement of Particles: Reply to Lowe', *Analysis*, 48: 65–72.

——(1993). 'Many, But Almost One', in K. Campbell, J. Bacon, and L. Reinhardt (eds.), *Ontology, Causality and Mind: Essays in Honour of D. M. Armstrong*. Cambridge: Cambridge University Press, 23–38.

——(1994). 'Humean Supervenience Debugged', *Mind*, 103: 473–90.

——(1999*a*). *Papers in Metaphysics and Epistemology*. Cambridge: Cambridge University Press.

——(1999*b*). 'Zimmerman and the Spinning Sphere', *Australasian Journal of Philosophy*, 77: 209–12.

——(2002). 'Tensing the Copula', *Mind*, 111: 1–14.

——(2004). 'Tensed Quantifiers', in D. Zimmerman (ed.), *Oxford Studies in Metaphysics*, i. Oxford: Oxford University Press, 3–14.

Lowe, E. J. (1987). 'Lewis on Perdurance versus Endurance', *Analysis*, 47: 152–4.

——(1988). 'The Problems of Intrinsic Change: Rejoinder to Lewis', *Analysis*, 48: 72–7.

——(2002). *A Survey of Metaphysics*. Oxford: Oxford University Press.

Ludlow, Peter (1999). *Semantics, Tense, and Time: An Essay in the Metaphysics of Natural Language*. Cambridge, MA: MIT Press.

——(2004). 'Presentism, Triviality, and the Varieties of Tensism', in D. Zimmerman (ed.), *Oxford Studies in Metaphysics*, i. Oxford: Oxford University Press, 21–36.

McDaniel, Kris (2003). 'No Paradox of Multi-Location', *Analysis*, 63: 309–11.

McTaggart, J. M. E. (1908). 'The Unreality of Time', *Mind*, 17: 457–74.

Markosian, Ned (1994). 'The 3D/4D Controversy and Non-Present Objects', *Philosophical Papers*, 23: 243–9.
—— (2001). 'Critical Study of Robin LePoidevin (ed.), *Questions of Time and Tense* (Oxford: Oxford University Press)', *Noûs*, 35: 616–29.
—— (2004). 'A Defense of Presentism', in D. Zimmerman (ed.), *Oxford Studies in Metaphysics*, i. Oxford: Oxford University Press, 47–82.
Mellor, D. H. (1981). *Real Time*. Cambridge: Cambridge University Press.
—— (1991*a*). *Matters of Metaphysics*. Cambridge: Cambridge University Press.
—— (1991*b*). 'Properties and Predicates', in Mellor (1991*a*), 170–82.
—— (1992). 'There are No Conjunctive Universals', *Analysis*, 52: 97–103.
—— (1998). *Real Time II*. London: Routledge.
—— and Oliver, Alex (eds), 1997. *Properties*. Oxford: Oxford University Press.
Merricks, Trenton (1994). 'Endurance and Indiscernibility', *Journal of Philosophy*, 91: 165–84.
—— (1995). 'On the Incompatibility of Enduring and Perduring Entities', *Mind*, 104: 523–31.
—— (1999). 'Persistence, Parts, and Presentism', *Noûs*, 33: 421–38.
Minkowski, Hermann (1909). 'Raum und Zeit', *Physikalische Zeitschrift*, 10: 104–11.
Noonan, Harold (1985). 'A Note on Temporal Parts', *Analysis*, 45: 151–2.
Oaklander, Nathan L. (1991). 'A Defence of the New Tenseless Theory of Time', *Philosophical Quarterly*, 41: 26–38.
——, and Smith, Quentin (1994). *The New Theory of Time*. New Haven and London: Yale University Press.
Oderberg, David (1993). *The Metaphysics of Identity over Time*. New York: St Martin's Press.
Oliver, Alex (1996). 'The Metaphysics of Properties', *Mind*, 105: 1–80.
Parfit, Derek (1975). 'Personal Identity', in Perry (1975), 199–223.
Parsons, Terence (1990). *Events in the Semantics of English: A Study of Subatomic Semantics*. Cambridge, MA: MIT Press.
Perry, John (1975) (ed.). *Personal Identity*. Berkeley and Los Angeles: University of California Press.
—— (1977). 'Frege on Demonstratives', *Philosophical Review*, 86: 474–97.
—— (1979). 'The Problem of the Essential Indexical', *Noûs*, 13: 3–21.
Pianesi, Fabio, and Varzi, Achille C. (2000). 'Events and Event Talk: An Introduction', in J. Higginbotham, F. Pianesi, and A. C. Varzi (eds.), *Speaking of Events*. Oxford: Oxford University Press, 3–47.
Plantinga, Alvin (1974). *The Nature of Necessity*. Oxford: Oxford University Press.
—— (1983). 'On Existentialism', *Philosophical Studies*, 44: 1–21.
—— (1985). 'Reply to my Colleagues', in J. E. Tomberlin and P. van Inwagen (eds.), *Alvin Plantinga*. Dordrecht: D. Reidel, 313–96.
Pollock, John L. (1985). 'Plantinga on Possible Worlds', in J. E. Tomberlin and P. van Inwagen (eds.), *Alvin Plantinga*. Dordrecht: D. Reidel, 121–44.
Prior, A. N. (1957). *Time and Modality*. Oxford: Oxford University Press.

Prior, A. N. (1959). 'Thank Goodness That's Over', *Philosophy*, 34: 12–17.
—— (1967). *Past, Present, and Future*. Oxford: Oxford University Press.
—— (1968*a*). 'Changes in Events and Changes in Things', in Prior (1968*c*), 1–14.
—— (1968*b*). 'Quasi-Propositions and Quasi-Individuals', in Prior (1968*c*), 135–44.
—— (1968*c*). *Papers on Time and Tense*. Oxford: Oxford University Press.
—— (1970). 'The Notion of the Present', *Studium Generale*, 23: 245–8.
——, and Fine, Kit (1977). *Worlds, Times and Selves*. London: Duckworth.
Quine, W. V. O. (1960). *Word and Object*. Cambridge, MA: MIT Press.
—— (1963). 'Identity, Ostension, and Hypostasis', in his *From a Logical Point of View*. New York: Harper & Row, 65–79.
Ramsey, F. P. (1925). 'Universals', in D. H. Mellor (ed.), *F. P. Ramsey: Philosophical Papers*. Cambridge: Cambridge University Press, 1980, 8–33. Reprinted in Mellor and Oliver (1997), 57–73.
Rea, Michael (1998). 'Temporal Parts Unmotivated', *Philosophical Review*, 107: 225–60.
—— (2000). 'Constitution and Kind Membership', *Philosophical Studies*, 97: 169–93.
—— (2003). 'Four-Dimensionalism', in Michael J. Loux and Dean W. Zimmerman (eds.), *The Oxford Handbook of Metaphysics*. Oxford: Oxford University Press, 246–80.
Reichenbach, Hans (1947). *Elements of Symbolic Logic*. London: Macmillan.
Robinson, Denis (1989). 'Matter, Motion and Humean Supervenience', *Australasian Journal of Philosophy*, 67: 349–409.
Rodriguez-Pereyra, Gonzalo (2003). 'What Is Wrong with the Relational Theory of Change?', in G. Rodriguez-Pereyra and H. Lillehammer (eds.), *Real Metaphysics*. London: Routledge, 184–95.
Russell, Betrand (1906). 'Critical Notice of *Symbolic Logic and its Applications* by Hugh MacColl', *Mind*, 15: 255–60.
—— (1915). 'On the Experience of Time', *Monist*, 25: 212–33.
Sattig, Thomas (2002). 'Temporal Parts and Complex Predicates', *Proceedings of the Aristotelian Society*, 102: 279–86.
—— (2003). 'Temporal Predication with Temporal Parts and Temporal Counterparts', *Australasian Journal of Philosophy*, 81: 355–68.
Schlesinger, George (1980). *Aspects of Time*. Indianapolis: Hackett Publishing Co.
Sider, Theodore (1996). 'All the World's a Stage', *Australasian Journal of Philosophy*, 74: 433–53.
—— (1997). 'Four-Dimensionalism', *Philosophical Review*, 106: 197–321.
—— (1999). 'Presentism and Ontological Commitment', *Journal of Philosophy*, 96: 325–47.
—— (2000). 'The Stage View and Temporary Intrinsics', *Analysis*, 60: 84–8.
—— (2001). *Four-Dimensionalism: An Ontology of Persistence and Time*. Oxford: Oxford University Press.
Simons, Peter (1987). *Parts: A Study in Ontology*. Oxford: Oxford University Press.

—— (1994). 'Particulars in Particular Clothing: Three Trope Theories of Substance', *Philosophy and Phenomenological Research*, 54: 553–75.

Sklar, Lawrence (1974). *Space, Time, and Spacetime*. Berkeley and Los Angeles: University of California Press.

Smart, J. J. C. (1962). 'Tensed Statements', *Philosophical Quarterly*, 12: 264–5.

—— (1963). 'The River of Time', in A. Flew (ed.), *Essays in Conceptual Analysis*. London: Routledge & Kegan Paul, 213–27.

—— (1980). 'Time and Becoming', in P. van Inwagen (ed.), *Time and Cause*. Dordrecht: D. Reidel, 3–15.

Smith, Quentin (1993). *Language and Time*. New York: Oxford University Press.

Taylor, Richard (1955). 'Spatial and Temporal Analogies and the Concept of Identity', *Journal of Philosophy*, 52: 599–612.

Teller, Paul (1984). 'A Poor Man's Guide to Supervenience and Determination', *Southern Journal of Philosophy*, Supplement to volume 22: 137–62.

Thomson, Judith Jarvis (1983). 'Parthood and Identity across Time', *Journal of Philosophy*, 80: 201–20.

—— (1998). 'The Statue and the Clay', *Noûs*, 32: 149–73.

Tichy, Pavel (1980). 'The Transiency of Truth', *Theoria*, 46: 165–82.

Tooley, Michael (1997). *Time, Tense, and Causation*. Oxford: Oxford University Press.

Unger, Peter 1980. 'The Problem of the Many', in P. A. French, T. E. Uehling Jr., and H. K. Wettstein (eds.), *Midwest Studies in Philosophy*, v Minneapolis: University of Minnesota Press, 411–67.

van Inwagen, Peter (1981). 'The Doctrine of Arbitrary Undetached Parts', *Pacific Philosophical Quarterly*, 62: 123–37.

—— (1990). 'Four-Dimensional Objects', *Noûs*, 24: 145–55.

Wiggins, David (1968). 'On Being in the Same Place at the Same Time', *Philosophical Review*, 77: 90–5.

—— (1980). *Sameness and Substance*. Cambridge, MA: Harvard University Press.

—— (2001). *Sameness and Substance Renewed*. Cambridge: Cambridge University Press.

Williams, D. C. (1953). 'The Elements of Being', *Review of Metaphysics*, 7: 3–18.

Williamson, Timothy (1998). 'Bare Possibilia', *Erkenntnis*, 48: 257–73.

—— (2000). 'Existence and Contingency', *Proceedings of the Aristotelian Society*, 100: 117–39.

Zimmerman, Dean (1996). 'Persistence and Presentism', *Philosophical Papers*, 25: 115–26.

—— (1998*a*). 'Temporal Parts and Supervenient Causation: The Incompatibility of Two Humean Doctrines', *Australasian Journal of Philosophy*, 76: 265–88.

—— (1998*b*). 'Temporary Intrinsics and Presentism', in D. Zimmerman and P. van Inwagen (eds.), *Metaphysics: The Big Questions*. Cambridge, MA: Blackwell, 206–20.

—— (1999). 'One Really Big Liquid Sphere: Reply to Lewis', *Australasian Journal of Philosophy*, 77: 213–15.

Index